THE POWER OF
FASTERCISE

Using the New Science of
SIGNALING EXERCISE
to Get Surprisingly Fit
in Just a Few Minutes a Day

DENIS WILSON, MD

CHELSEA GREEN PUBLISHING

White River Junction, Vermont
London, UK

FASTERCISE is a trademark of Fastercise, LLC.

Project Manager: Alexander Bullett
Editor: Fern Marshall Bradley
Copy Editor: Deborah Heimann
Proofreader: Eliani Torres
Indexer: Linda Hallinger
Designer: Melissa Jacobson
Page Layout: Abrah Griggs

Printed in Canada.
First printing October 2019.
10 9 8 7 6 5 4 3 2 1   19 20 21 22 23

Our Commitment to Green Publishing

Chelsea Green sees publishing as a tool for cultural change and ecological stewardship. We strive to align our book manufacturing practices with our editorial mission and to reduce the impact of our business enterprise in the environment. We print our books and catalogs on chlorine-free recycled paper, using vegetable-based inks whenever possible. This book may cost slightly more because it was printed on paper that contains recycled fiber, and we hope you'll agree that it's worth it. *The Power of Fastercise* was printed on paper supplied by Marquis that is made of recycled materials and other controlled sources.

Library of Congress Cataloging-in-Publication Data
Names: Wilson, Denis, 1960- author.
Title: The power of fastercise : using the new science of signaling exercise to get surprisingly
    fit in just a few minutes a day / Denis Wilson.
Description: White River Junction, VT : Chelsea Green Publishing, [2019] | Includes
    bibliographical references and index.
Identifiers: LCCN 2019025544 (print) | LCCN 2019025545 (ebook) | ISBN 9781603588997
    (paperback) | ISBN 9781603589000 (ebook)
Subjects: LCSH: Weight loss. | Reducing diets. | Metabolism. | Nutrition. | Exercise.
Classification: LCC RM222.2 .W457 2019 (print) | LCC RM222.2 (ebook) | DDC 613.2/5—dc23
LC record available at https://lccn.loc.gov/2019025544
LC ebook record available at https://lccn.loc.gov/2019025545

Chelsea Green Publishing
85 North Main Street, Suite 120
White River Junction, VT 05001
(802) 295-6300
www.chelseagreen.com

r of Fastercise

"I have long been an avid follower of Dr. Wilson, and his innovations in thyroid management have informed my own practice since the late 1990s. Dr. Wilson has always been a trailblazer in medicine, a systems thinker, and proponent of whole body-mind-centered therapies. In a world where we are overwhelmed, overfed, and undernourished on all levels of our being, Dr. Wilson has created a way and convenient way to enhance metabolic flexibility at the cellular level. By retraining our signaling pathways with exercises and concepts that are both ancient and new and for the most part free and accessible by all, Dr. Wilson is once again creating an opportunity to reset what has become increasingly broken in our population, restoring us to optimal health."

—**Nasha Winters**, ND, FABNO,
coauthor of *The Metabolic Approach to Cancer*

"Dr. Denis Wilson led a renegade overhaul of the medical approach to treating thyroid imbalances, pioneering his effective clinical protocol known as Wilson's temperature syndrome. Now this innovative thinker has done it again with his logical and surprisingly simple approach to losing body fat and building muscle. Wilson's message in *The Power of Fastercise* is clear, his reasoning flawless, and his guidance is sure to help readers follow these easy guidelines to restoring or achieving a healthy weight."

—**Dr. Jill Stansbury**, ND, author of
Herbal Formularies for Health Professionals

"*The Power of Fastercise* puts forth a new and interesting theory about how small amounts of small and purposefully distributed movements could matter tremendously to our well-being."

—**Katy Bowman**,
best-selling author of *Move Your DNA*

"Denis Wilson's new book, *The Power of Fastercise*, is desperately needed. Conventional medicine has failed us. We are becoming fatter and sicker, and we need a new paradigm to improve our health. Exercise helps every person and can improve any condition. Dr. Wilson explains how anyone can implement his program to not only lose weight but to improve their immune function. I will recommend fastercise to my patients, and I highly recommend this book."

—**David Brownstein**, MD, medical director,
Center for Holistic Medicine; national best-selling author of health books
and *Dr. Brownstein's Natural Way to Health* newsletter

"Sometimes people come up with ideas of how we can get more done in less time with a quick and effective workout, but those ideas turn out to be just another regurgitation of programs and concepts that are already out there. *The Power of Fastercise* is very, very different. Dr. Denis Wilson has cracked a code on how we can signal our body to shed body fat and gain muscle with simple exercises that anyone can do. He shows us that how and when we exercise can make a dramatic difference. With fastercise you can lose weight without falling for fad diets or spending hours doing boring cardio."

—**Allan Misner**, author of *The Wellness Roadmap*; host, 40+ Fitness podcast

"Denis Wilson has taught thousands of doctors how to crack the code on weight loss. He has the unique ability to take the complicated metabolism story and make it achievable. Fastercise is the solution that so many of us need today. Not a gimmick or a shortcut, fastercise is based on tried-and-true principles of healthy eating and physical activity sprinkled with extra energy to enjoy. *The Power of Fastercise* is the perfect book for both expert doctors and clients looking for more vibrancy and success in reaching their health goals."

—**Mark Menolascino**, MD, MS, ABIHM, ABAARM, IFMCP, medical director, Meno Clinic—Center for Functional Medicine; author of *Heart Solution for Women*

"In this informative and valuable book, Dr. Denis Wilson explains the details of his fastercise framework of quick muscle activity, which provides the potential for a wide range of people to lose weight, build muscle, and boost metabolism."

—**Dr. Kevin Spelman**, CEO, Health, Education, and Research in Botanical Medicine

"Denis Wilson is a deep thinker who pays tremendous attention to detail. In *The Power of Fastercise*, he opens our minds to a new understanding of how the individual elements of metabolism come together to play a symphony—and gives us the wisdom to have it play the tune that we like. If you are interested to understand your body better—or just want a simple but profound plan of action to reach your weight and health goals—then this book is a must-read."

—**Gerrie Lindeque**, MD, founder, Made to Be Well

To the light and life of the world.
And to my wife, the mother of our
five children, Lisa. And to my esteemed
colleague Michaël Friedman. Thank you
all for the many years of love and support.
And to all those who are searching for
solutions to their health problems.

CONTENTS

Introduction

Many people would love to lose fat and build muscle. It sounds simple but a lot of us have found it extremely difficult. It's as though we want to get in great shape but our bodies are determined to hold on to that fat! Why? Imagine how much easier it would be to lose fat and keep it off if we could persuade our bodies to *want* to store less fat. Understanding the body's innate priorities can help us understand how to align our efforts so that our bodies work with us instead of against us. What a huge difference it can make when we use tools as they were designed rather than some other way. Imagine how hard it would be to turn a screw using the wrong end of a screwdriver.

The good news is that you may be able to lose fat and build muscle very effectively simply by sending your body the signals that will put its mind-boggling complexity to work for you. The science of signaling exercise is amazingly simple yet delivers such powerful results that it will liberate many people, empowering them to transform their lives and help them reach their potential. The strategies I present in this book to burn stored fat, build muscle, boost metabolism, and improve your overall health are so easy to integrate into your normal day that they are easy to sustain indefinitely. You may almost feel as though your body is changing automatically. These strategies don't require special equipment, expensive meal plans or workout clothes, or hours and hours of time every week. Your body's built-in mechanisms for burning stored fat will do the job for you.

I believe that looking at things differently can often help us see things differently and help us find solutions to our problems. About thirty years ago, I became particularly interested in the relationships among metabolism, body temperature, and thyroid function. Thyroid function largely determines metabolic rate and body temperature. I observed that many people suffer from low body temperatures (an average daily body temperature below 98.6°F [37.0°C]) and symptoms of slow metabolism such as fatigue, depression, easy weight gain, fluid retention, premenstrual syndrome, anxiety, decreased memory and concentration, and many others even though their blood levels of thyroid hormones are in the normal range as measured by standard blood tests. These symptoms often develop suddenly after a person encounters major physical, mental, or emotional stress. I identified this reversible form of slow metabolism as Wilson's temperature syndrome (WTS). I was the first doctor to ever use sustained-release T3 and developed a unique T3 therapy protocol that is very successful in normalizing body temperature and resolving the symptoms of slow metabolism. Many patients remain improved even after the treatment has been discontinued. I have written a book about WTS for patients and a treatment manual for doctors and developed an informational website, www.wilsonssyndrome .com. Over the years, I've trained thousands of doctors in WTS, and doctors are successfully treating people around the world. The treatment of WTS is also being taught in naturopathic medical schools.

Looking for New Answers

Many WTS sufferers tend to gain weight even though they don't feel they are overeating. When their temperatures are normalized with treatment, their symptoms often resolve and they're able to lose excess weight, but some people continue to have trouble losing weight even after their temperatures have been normalized and all their other symptoms have resolved.

Through my work with thousands of patients and doctors over the last three decades, I have seen how disheartening and frustrating it is for those who try very hard to lose weight but find that nothing seems to work for them. Indeed, their struggle seems to defy the laws of science. After all, how can people eat less, exercise more, and *still* gain weight?

One reason that people have so much trouble losing body fat is that much of what they've heard or read on the topic is either inaccurate or oversimplified. The human body is a complex system that adapts masterfully to its conditions. It's important to first understand how your body works in order to direct it to function the way you want it to. A surge of recent medical research has greatly expanded our understanding of the inner workings of the body. In this book, I explore how this research and this new understanding can open the door to easily sustainable fat loss for many people.

Without this new understanding to guide us, following the usual advice of "eat less and exercise more" can easily lead to disappointing results. There are so many diet and exercise approaches promoted through books, websites, and weight-loss centers that seem to contradict one another, so it can be very hard to know which are based on sound science and might provide lasting results.

On top of the disappointment and frustration of following expert advice to the letter and making little to no progress, people who are overfat can often face the added insult of being accused of not following the given program. They can also face judgment and stigma from people in general. People who haven't struggled with weight issues may judge overfat people to lack interest, ability, determination, or even value. In the face of so much condemnation, overfat people may even judge themselves or question their own sanity. They can have feelings of frustration, disappointment, discouragement, inadequacy, anger, defensiveness, depression, anxiety, guilt, and low self-esteem. Even small tasks such as changing a diaper or taking out the trash can seem impossible. After all, they may think, what's the use? "Nothing works for me." By the way, there are a lot of physiological changes that can be associated with being overfat that can contribute to such feelings. All this can lead people to find comfort in food, which can make matters worse in the long run.

I believe that there is always a solution to our problems and that we just need to keep looking to find the solution that works for us. Thank goodness for the internet, which now makes this easier than ever. The worldwide rise in obesity is a very serious problem that has proved to be difficult to solve. Yet some people are very lean and muscular, so there has to be a way to help some overfat people to become more lean.

Whenever we try to fix something that's not working, whether it's a kitchen appliance, a toilet, or the human body, it's really helpful to know or find out how it's built and how it's supposed to work. That knowledge can provide clues on what we can do to restore proper function. So in my effort to help people struggling with excess weight, I started back at the beginning, reviewing the basics about how the body is built and how it works. It was an eye-opening experience.

Two aspects of my process stand out. For one, when I was in college and medical school, we usually studied one subject at a time, and for each subject—cardiology, gastroenterology, endocrinology—there was a separate textbook. But for my recent review of the human body, I found the internet an amazing resource. It is like one immense book that can be used to study all imaginable topics at the same time. That ability to compare information across disciplines provides new ways to connect the dots and reach new understanding. The other aspect is how much the understanding of and information about the human body has changed and expanded since I went to medical school. I could never have imagined it.

This wealth of information gives us the opportunity to rethink all the assumptions that we've made and have been taught. The old ideas were drawn from an older, smaller set of information. A much larger set of new information offers an opportunity to come up with new methods and conclusions. All we have to do is load up a new set of data and let it percolate in our minds and see what new connections emerge.

In my review of basic science, I learned some things I had never known before and relearned some things I had forgotten. There have been many "aha" moments along the way. The first of these came after I had been studying for many days ATP, mitochondria, energy storage, muscle function, the Krebs cycle, appetite, fasting, high-intensity exercise, hormones, and more. One night, I went to sleep with all those things on my mind, trying to understand them and how they fit together. The next morning when I woke up, I spontaneously tightened my muscles in a "morning stretch" as I always do. As I enjoyed that good feeling, I found myself wondering *why* we have the natural inclination to contract our muscles hard like that every morning as soon as we wake up. I hypothesized that it was to help us mobilize the *stored* energy that would help us start the day, since our bodies had gone without food for many hours while we slept. If that were so, I mused, then we ought to be able to contract our muscles hard for a few seconds at *any* time of day to mobilize some stored energy, which might help to hold hunger pangs at bay and allow us to better control our eating choices. In that moment, I began to build my concept of "fastercise." Fastercise is the coordination of diet and exercise on a *meal-by-meal basis according to appetite* to incline the body to store less body fat and build more muscle. Fastercise is the first program to *manage appetite with exercise* and manage exercise according to appetite.

Understanding the Body's Signals

Sometimes it can feel as though our bodies are fighting against us. But maybe we're just having a big misunderstanding with our bodies. Some fitness problems boil down to simple miscommunication. In recent medical literature, terms such as *signaling*, *perception*, and *response* are showing up more and more. This research shows that our bodies respond to signals they perceive. Wow. That's really incredible when we think about it. Given the right signals and resources, the human body can grow, heal, think, sing, dance, paint, run, jump, and reproduce. The human body is incredibly complex, yet it is built to respond to simple signals to help make our lives as easy, successful, and pleasant as possible, *automatically*.

Automation is at the base of the modern technology that has revolutionized many aspects of our lives. Like the human body, machines and computers respond to signals they perceive. "If this, then that," is a phrase at the heart of computer programming. If this signal is perceived, then that action is taken. This automation helps run the internet, cell phones, cruise control, air travel, space exploration, an endless array of home appliances and fascinating entertainment, and many other conveniences. If we send the right signals by pressing the right buttons in the right sequence at the right times under the right conditions, then wonderful things can happen. However, if we *don't* send the right signals, then none of the magic of modern technology happens, which can be disheartening and frustrating.

The same can be said for the human body. In fact, I contend that the complexity of the automation happening in your body on a daily basis far outstrips the technological "miracles" of computers, smartphones, and smart homes. There are so many amazing things that our bodies can do automatically. They clean our blood, make sure we get enough oxygen, digest food, refresh us with sleep, and maintain our body temperature, to name just a few. Sometimes, though, it can seem that our bodies also gain fat automatically. Imagine if we could send our bodies simple signals that directed them to automatically lose fat and build muscle instead. Fortunately, that is possible, and in this book, I explain in detail both the science and the practical actions you can take to send such signals to your body.

There's certainly been a disheartening and frustrating rise in obesity in the United States. In the last forty years, the percentage of people who are obese has ballooned, even among children. In the United States today, more than 35 percent of adults and nearly 17 percent of children between the ages of two and nineteen are obese. Obesity is known to contribute to autoimmune and chronic diseases such as diabetes, hypertension, heart disease, stroke, dementia, and many others. As worrisome as these medical implications are, many people find the day-to-day physical and social inconveniences of obesity even more troubling. People want to be able to fit into their clothes comfortably, to move comfortably, and to enjoy social interaction comfortably. When our weight is out of control it can feel like our lives are out of control.

It is likely that many factors contribute to the current obesity epidemic. Perhaps the most significant factor is the signals we send our bodies, particularly the signals we send through the composition and timing of our diet and physical activity. By working backward from what new research reveals about the way the body functions, I identified a new set of signals—the fastercise techniques of tightercise and shivercise—that encourages the body to burn fat and build muscle.

I've seen great results with fastercise, and so has my daughter Allison. In my case, when I started fastercising in the spring of 2017, my body mass index (used as an estimate of obesity) was 23.9, which put me in the high end of the healthy range for a man. After a couple of months of fastercise, I had dropped about twenty pounds and my body mass index had shifted to 21, which is in the middle of the healthy range. Even though I wasn't in the obese range before starting fastercise, I was surprised at how much better I felt as my body changed. For example, I found that I could move around a lot more easily and that I had greater range of motion. I no longer experienced acid reflux. And if I bent over to pick up something off the floor, my face didn't remain red so long afterward. (I now think that the two latter symptoms may be related to excess fat around the internal organs.) I've also enjoyed more energy and enthusiasm, sharper thinking and better focus, more rejuvenating sleep, and greater resistance to illness. I also get far less winded now when I exercise, I feel stronger, and my skin has a more youthful appearance.

In Allison's case, she had run track and cross country in high school, working out for four hours most days. She said she felt invincible. Then she got married, graduated college, and last year she had a baby. She was left with thirty extra pounds postdelivery, she didn't fit into any of her regular clothes, she had less range of motion in her movements and felt sluggish. When she looked in the mirror, she was so disheartened by the appearance of her body. However, she had almost no free time to devote to fitness training. She had recently moved across the country and started a new full-time job. She was taking care of a newborn baby and working on launching a social media business and buying a house! But one day she noticed how much more fit I was looking and asked me what my secret was. I told her about fastercise, and she was intrigued. Fitness in just a few minutes a day? So Allison started daily fastercise. She was able to easily incorporate it into her workday in a professional office setting at a Fortune 15 company. Now she reports: "I dropped thirty-five pounds over the course of three months. I weigh even less than my pre-pregnancy weight. I also built more muscle, both in size and definition, than I've ever had." For Allison, the best part of fastercise is that she spent less than ten minutes a day exercising, splitting it up into one- and two-minute increments. When she ate a meal, she could enjoy every bite. "I feel fantastic and look fantastic! I even fit back into my high school clothes!" Allison says. She didn't think it would be possible to lose weight given her job, her family, and other commitments, much less to lose it so quickly. But the pounds flew off and stayed off. Allison reports that she even took a break from fastercising over the holidays, with no rebound in her weight.

I have led hundreds of doctors in one minute of shivercise in my presentations at medical conferences, and many of them were very excited and hopeful about the effects. I've received comments such as "I'm excited not only for my patients but also for myself. I'll finally be able to get fit even though I don't have time available to exercise!" The enthusiastic response from my colleagues confirmed that fastercise would be embraced by busy people who want to achieve extraordinary fitness. It motivated me to continue to research the science and develop the concept of fastercise, and to present all that I have learned, experimented with, and refined in the pages of this book.

There are many obstacles that can make it difficult to get fit. Two of the biggest reasons people give for not getting in shape is that they don't have the time and they don't like being hungry. When we push off hunger with fastercise we save time because there's no cooking, eating, or digestion involved, and no dishes to clean! And fastercise can eliminate hunger within eight to sixty seconds. Perfect. Another obstacle to getting in shape is not having or wanting to spend a lot of money on special diets, programs, gym memberships, personal trainers, and personal chefs. Some celebrities and athletes get paid to get in shape and they have the time and money to do so. People who are able to build muscle and lose fat are often given a lot of credit and honor and glory, while people who don't have that opportunity often face discrimination, with obese women facing more discrimination than obese men.[1] I feel the information in this book completely levels the playing field when it comes to fitness and puts extraordinary fitness within easy reach of almost everyone. Gaining control of our bodies can give us better control of our lives. We can even use this information to get "paid" to lose fat just like a celebrity because food is one of our biggest expenses. Every time we push off hunger with fastercise and burn stored fuel instead of eating, we save money on food. Money saved is money earned.

Quite frankly, I have found the ease, simplicity, and effectiveness of signaling exercise absolutely flabbergasting, especially for busy people who don't have the time, money, ability, or desire to spend hours in the gym or grocery store. In fact, by using the solutions I present, many people will be able to spend less time and money losing fat than they're currently spending gaining fat! It's astonishing how easy it can be to lose fat and gain muscle, strength, resources, and control when we work with the body rather than against it. The science and application of fastercise put fantastic fitness within easy reach of almost everyone: seasoned athletes, fitness enthusiasts, and even those who dislike exercising or have physical limitations. Whatever your fitness goals are, fastercise can help you achieve them.

CHAPTER 1

The New Science of Signaling Exercise

D o you have a cell phone? If you do, chances are that you interact with your phone many times a day—to tell it to make a call, send a text or a tweet, or check the weather or see what your friends have posted. Your phone interacts with you by making sounds, vibrating, and displaying information on the screen for you to see. You interact with your phone by touching the screen and pressing buttons. In other words, your phone sends you signals, and you respond by sending your phone signals. Your interaction with your body is very similar. Your body definitely sends you signals, and you can respond by sending signals to your body. One of the most basic signals that the body sends is hunger. Hunger is the signal the body sends to let you know that the fuel it's currently burning is running low. If you eat, your body will burn the food you consume to provide the energy it needs. But if you don't eat, your body will begin to burn stored glycogen and fat and also muscle tissue. That's not a good thing, because we want to preserve our muscle. What if we could direct the body to burn only stored fat instead of muscle? The remarkable fact is that you *can* direct your body to burn stored fat and to build muscle, simply by doing a very short burst (three to sixty seconds) of specialized signaling exercise I call fastercise. As you fastercise, your body increases its burning of stored fat and your hunger goes away in moments. It's as simple as that. Seriously? Seriously.

Diet and exercise send powerful signals to our bodies, and can promote optimal health. However, when we think of diet and exercise as being only about our intake and expenditure of calories, we can get some very confusing results: "Why am I gaining weight even though I'm eating less and exercising more?" It's important to take a broader view of the signals we send the body with diet and exercise. In my opinion, the human body is the most advanced technology on earth, and given the right signals, circumstances, and resources, our bodies can do incredible things automatically.

What Is Fastercise?

Fastercising between meals is a super convenient, natural, and effective way to fit in a healthy, sustainable, and enjoyable lifestyle. Fastercise is a form of high-intensity exercise that is usually done for anywhere from a matter of seconds up to about a minute or two. Fastercise is based on two body movements that are so natural our bodies do them automatically—the "morning stretch" (which is actually a period of tightening up various muscle groups) and shivering (what we do when we are stuck in a cold place for too long). I call the two variations of fastercise that are based on these two movements *tightercise* and *shivercise* because they are good descriptions that are easy to remember.

Tightercise and shivercise are not difficult to do. You may feel a little awkward the first couple of times you try them since they are so unconventional, but before you know it, they will become so natural that you may even find yourself doing them automatically.

Tightercise

Tightercise is simply the act of voluntarily contracting your muscles as hard as you can for three to eight seconds (or longer), the way you do when you wake up and stretch in the morning. Why do we do that every morning? I believe that one of the most important reasons is that our bodies are low on fuel when we first wake up, because we haven't eaten for hours. Contracting our muscles really hard for a few seconds mobilizes our energy stores to push off our hunger and give us the energy we need to get up, hunt for breakfast, and start our day. As it turns out, contracting our muscles this way is useful whenever we are low on fuel, not just when we wake up. For example, later in the morning when we get hungry for a snack, we can tightercise for several seconds to instantly push off our hunger for a time. And we can repeat that a number of times if we wish.

Tightercising can also help mobilize fuel for other purposes, too, such as when we are about to expend a lot of energy in an athletic event.

In technical terms, tightercise is an isometric resistance exercise. You can contract several muscles at a time for three to eight seconds, and can run through all of your big muscle groups in about a minute. Or you can tighten up all of your muscles at once and be done in eight or ten seconds.

First thing in the morning is a good time to tightercise because the body is naturally inclined to contract muscles automatically upon waking. A good sign that you have tightened enough of your muscles hard enough for long enough is that you become winded enough to catch a deep breath. I explain the significance of this deep breath in greater detail later in the book, but briefly stated, breathing is one mechanism that the body uses to get rid of acid that can build up in muscle tissue with exercise. The faster you exercise, the more acid can build up, which signals your body to increase respiration. You'll find it's not difficult to catch a deep breath after tightercising, because there's a natural tendency to breathe less while you contract your muscles. You may experience a deep, full inhalation and exhalation within ninety seconds after you tightercise. This deep breath is similar to a full-fledged yawn. The muscle contractions and deep breaths we experience by tightercising enable our bodies to mobilize and burn energy stores to generate the energy we need to get a great start on the day.

Shivercise

Shivercise is the act of intentionally rapidly shifting between contracting (tightening) and relaxing your muscles, causing them to shiver, shake, or shudder. These small-amplitude movements (less than a couple of inches) are performed as fast as possible. Picture shivering or running in place while moving your arms and legs only a couple of inches in either direction, lifting only your heels off the ground. The purpose is to maximize the number of muscle contractions and relaxations per minute. Speed is the key. Though we can shivercise when we feel cold and want to feel warm, we can also use it to become winded enough to catch a deep breath.

There are similarities and differences in the signals that tightercise and shivercise send your body. Both signal the body to mobilize stored energy (in the form of glycogen and fat) in order to provide the body with instant energy, but shivercise does this more aggressively. Both techniques can also stimulate muscle growth, but tightercise more so than shivercise. Does being able to lose fat and preserve muscle without much discomfort, time, money, or slowing of the metabolism seem too good to be true? Actually, canceling hunger with fastercise often does deliver these benefits.

It can be very difficult to motivate ourselves to get up, get dressed, and drive across town for a strenuous workout. But with fastercise, we just get started, contract our muscles for a minute or so, catch a deep breath, and we're done. By the time it starts getting difficult, we're done. Fastercise is exercise that almost anyone can do; after all, almost anyone can shiver or tighten some muscles, right? This includes older people, people with arthritis, and postpartum women. In the introduction, I described my daughter Allison's experience. She was advised to avoid strenuous workouts for six weeks postpartum. In the meantime, she was very happy with the benefits she saw with the fastercise that she was easily able to do.

Through the fastercise techniques of tightercise and shivercise, we can send more powerful signals to our bodies in just a couple of minutes than we might otherwise accomplish in an entire day. To get the most benefit from fastercise, I also recommend that people drink plenty of water. Taking supplements of vitamins, electrolytes, and branched-chain amino acids (BCAA) can also be helpful. In chapter 6, I discuss in detail how these measures support your efforts to burn fat and build muscle.

The timing of signaling exercise in relation to eating is also a key to success. By choosing the right timing of fastercise in relationship to eating, we send powerful signals to the body that direct it to work with our efforts rather than against them. To understand how this works, let's start by learning more about the body's prime directive: survival.

Understanding the Body's Priorities

The basis of all good communication is mutual understanding, which includes understanding one another's priorities. Fortunately, the body's priorities are quite easy to understand, and learning about them helps to explain many mysteries about fat loss and muscle building that are otherwise confounding.

When hunger arises, your body has the following priorities, listed here from most urgent to least:

1. Avoid passing out in this immediate moment.
2. Avoid being killed by a predator, an enemy, or a dangerous situation in the next few minutes.
3. Avoid damaging your muscles.
4. Avoid starving during the next several months if food becomes scarce.

These priorities may not seem to fit many of us in this day and age; however, for our ancestors finding enough food for survival was a daily struggle, and the risk of being killed by predators and enemies could have been very high. These priorities have been hard-wired into our bodies for countless generations.

For modern humans, the risks are different. Falling asleep at the wheel can easily have fatal consequences literally in the blink of an eye. The same outcome can occur if the body fails to supply the brain with sufficient energy—that is, you might pass out. The brain relies heavily on burning glucose for energy. That's why providing sufficient glucose to the brain at all times is the body's highest priority when it comes to survival. Similarly, surviving an attack or dangerous situation may depend entirely on our bodies being able to deliver sufficient fuel to our brains and muscles to give us the sudden rush of energy we need to fight or flee. The ability to fight or flee is also why it's crucial for our bodies to preserve our muscles. Finally, our bodies need to do what they can to survive as long as possible if food becomes scarce.

Our bodies work to balance these four priorities as well as they can, considering the priorities sometimes conflict with one another. For example, muscles are super important for the survival of the body. Why would the body ever degrade muscle? It would because brain function is even more important for survival than muscles are. Our brains need a steady supply of glucose in order to function properly. You may not think of muscle as an energy-storage site, and muscles do serve other critical functions, but in some circumstances, the body will break down muscle tissue and other proteins for use as fuel to provide energy. Degrading a little muscle when blood glucose is low can provide additional glucose so the brain can keep functioning. This addresses priority number one: Don't pass out, because otherwise you may end up getting into a life-threatening accident.

Here's another example. Priority number four is to save as much stored energy as possible, because that would allow you to remain alive as long as possible in times when there is little or no food. Why would your body choose to burn up that stored energy? It would do so to help you escape from a dangerous situation, such as getting out of a burning building or jumping out of the path of a speeding car. Having a big supply of stored energy won't do you any good if you die in the next sixty seconds! And in times past, humans sometimes needed to expend stored energy to chase down prey or go foraging in order to obtain food to avoid the danger of starvation.

With these four universal priorities of the body in mind, we can better understand the signals the body is sending us and the signals we can send back to the body to direct it to help us accomplish our health and fitness goals.

Hunger Starts as a Question, Not a Demand

As the energy from your last meal runs out, the hunger you begin to feel is your body's way of asking you what fuel you'd like to burn: food, or biomolecules already stored in the body. At that point, if you don't eat again, your body begins to burn glycogen (a type of carbohydrate that is stored in the liver and muscles), muscle, and fat to access the energy stored there. The longer you go without food and the more your body's energy stores are depleted, the more hunger becomes a demand. Thus, when you eat you build glycogen, muscle, and fat, and when you don't eat you burn glycogen, muscle, and fat.

When we eat, we tend to gain fat as well as muscle and when we don't eat we tend to lose muscle as well as fat. However, a desirable goal is to *lose fat* and *gain muscle*, or at least preserve muscle, in order to feel better and move with greater ease. Trying to lose fat and gain muscle can be like trying to unlock a padlock. Unless you know the right combination, you aren't likely to succeed. It is possible to lose fat and gain muscle—I'm sure you've seen before-and-after pictures of people in magazines, advertisements, or fitness blogs who have done it. Apparently, there is a particular combination of factors and priorities that unlock our fat stores for removal. Fortunately, we can align ourselves with the body's priorities and use fastercise when hunger arises to signal the body to unlock our fat stores.

When your body signals you with hunger pangs, asking, "Where would you like to get your fuel?" you can respond in one of three ways.

- You can ignore it: Don't eat and don't fastercise.
- You can fastercise instead of eating.
- You can eat.

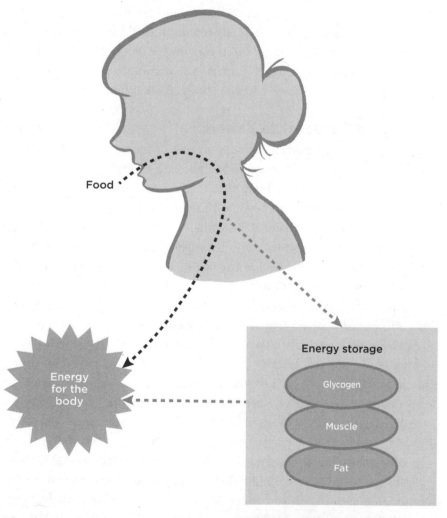

Figure 1.1. When you eat, your body metabolizes that food into energy that is available right away. If there is extra energy that isn't needed at the moment, the body stores it in the form of glycogen, muscle, and fat.

Ignoring Hunger Pangs

When you ignore hunger, neither eating nor fastercising, you are telling your body that there is no food available—and your body interprets that as an increased risk of starvation, should the lack of food continue for a long time. Since avoiding starvation is one of the body's four inherent priorities, it gets the message to burn stored fuel to power the body and to hunt for more food, but it also gets the message to conserve energy. When you abstain from eating, your hunger will come and go. Each time your body burns enough glycogen and muscle to provide glucose for the brain, hunger goes away. When that glucose is used up, hunger pangs return as a signal that the body is burning muscle again. As you continue to go without food, your metabolism slows down and your body temperature drops.

Answering Hunger with Fastercise

Fastercising gives the body a different answer to the hunger signal. When you answer hunger by fastercising strenuously enough to catch a deep breath, you are telling your body that there is food available, and you're seeking it out. You need the body to burn stored fuel to provide you the energy you need to succeed in hunting or foraging. This is your body making an energy investment in a promising opportunity in the hopes of obtaining more energy, akin to the old saying, "It takes money to make money." At the same time, your body is receiving the message that you need your muscles to be successful at obtaining food, and in response, your body will work to preserve and build those muscles.

In addition, fastercise simulates escaping a dangerous situation, which is an even higher survival priority than avoiding starvation. Swift and strong muscle function is critical for you to successfully obtain food or escape from a dangerous situation. Fastercise sends the body the message to burn stored glycogen and fat to provide the energy you need and to preserve or build muscle to provide the strength and speed you need. Your body responds by burning stored fuel, so it doesn't need any additional fuel—it doesn't need to burn muscle and it doesn't need more food. That's why your hunger goes away when you fastercise. Have you ever noticed that your hunger often goes away while you are exercising? Your body is burning stored fat and glycogen, using (preserving) your muscles, and you don't feel hungry. This is one reason I believe that hunger is a sign that the body

Table 1.1. Signals Summary

Signal Sent by the Body	Meaning of Signal
Hunger pangs	Body is burning muscle, metabolism is slowing.
Hunger abates after fastercise	Body is no longer burning muscle; it is burning glycogen and stored fat.
Hunger abates after eating	Body is no longer burning muscle; it may be storing fat.
Hunger comes back aggressively	Body is running out of fuel quickly; it is burning muscle.
Deep breath following fastercise	Body has been stimulated to produce more mitochondria.
Involuntary stretching	Body is mobilizing stored fuel.
Feeling poorly (for those on a low-carb diet)	Body likely needs water and electrolytes.

Signal Sent to the Body	Meaning of Signal
Fastercising when hungry	Keep and build muscles; burn glycogen and stored fat to ease hunger. Reduce tendency to store fat.
Not eating or fastercising when hungry	Burn stored fuel, slow metabolic rate; burn muscle to supply glucose for brain. Increase tendency to store fat.
Eating without fastercising when hungry	Food is scarce. Preserve and build fat stores.
Fastercising before eating	Food is plentiful. There is little need to store fat.
Eating lightly (only enough to satisfy appetite)	Food is plentiful. There is little need to store fat.
Overeating (eating more than needed to satisfy appetite)	Food is scarce. Preserve and build fat stores by eating a lot if you get the chance.
Eating when not hungry	Food is scarce. Store fat, make less muscle.
Shivercising	Burn glycogen and stored fat; preserve muscle; build fat-burning machinery.
Tightercising	Burn glycogen and stored fat; preserve muscle; build muscle size and strength.
Fastercising when feeling mentally fatigued	Wake up, pay attention, focus, learn, and work.

is burning muscle. Burning muscle is a more serious matter than burning glycogen and fat, and so the body notifies us about it.

With some experimentation, you'll discover that you can cancel your appetite faster by fastercising than by eating. That's because stored fuel is right there in the body's tissues, ready to be released immediately, whereas food needs to be chewed, swallowed, and digested before the body can access the energy it provides. And, as an additional benefit, fastercise encourages your metabolism to speed up rather than slow down, so when you lose fat by fastercising, it's easier to keep that fat off. (I explain more about how this works in chapter 6.)

Answering Hunger by Eating

When you answer hunger by eating, you are telling the body to burn food for fuel instead of using up your stores. But when you *overeat*, the message you send is slightly different. You are signaling your body that there may not be food available tomorrow or the next day. In effect, you are telling your body to store as much fat as possible to prepare for imminent deprivation. Your body responds by storing more fat and making less muscle, and you may experience cravings for high-energy foods.

Thus, feeling hungry is a valuable signal. Pay close attention to when and how hunger arises during the day, and also to when it goes away. When it does go away, it's important to stop eating if you want to lose fat and build muscle.

When I change my diet, I notice that the kinds of foods I crave when I get hungry also start to change. I tend to get hungry for some of the new dietary items I've started eating. Similarly, when my eating schedule changes, I tend to get hungry at the new mealtimes. I think these tendencies we have as individuals help shape the development of the cuisines of different cultures. A group of people start enjoying certain foods and they develop cravings for those familiar foods, so they tend to keep eating them, and figuring out new and varied ways to prepare them. Whatever we do trains our bodies, and our bodies get good at responding to the training, whether that's gaining fat and losing muscle or gaining muscle and losing fat. Thus, we need to be careful what habits we develop, because every habit sends a signal to the body. Pay close attention to the signals your body sends you as well. Although we are all similar, we can have different circumstances, preferences, and goals. There is no "one size fits all" diet and exercise program, but I think you'll find that you can adapt fastercise to suit your needs and preferences quite easily and successfully. As long as you speak its language, your body will cheerfully follow your instructions!

The differences in how the body responds to each of the various signals we can send when we're hungry begin to reveal the real power of fastercise. As I noted earlier in this chapter, it's incredible to think of all the functions that the body does automatically. We don't need to think about breathing, or digesting food, or pumping blood, or making urine. However, when we have to take over any of those processes and do them manually (that is, with respirators, dialysis, hormone therapy, or blood sugar management) we immediately discover how complicated they are. Even with all the sophisticated medical technology we've developed, our bodies can do a better job naturally (when they are healthy).

Fastercise is powerful because it works *with* the body's incredible automatic metabolic processes. With very little time and effort, we can time our meals and fastercise so that our fat burning and muscle building encourage one another rather than fight against each other. Think of the reminders, alerts, and notifications we receive by text, email, or alarm that prompt us to do tasks at the right time. The problem is that we don't always see or hear such electronic reminders. However, hunger is pretty hard to miss. Hunger has a way of getting our attention. Day in and day out, hunger will remind you to cancel hunger with fastercise, which will enable you to lose fat and gain muscle automatically. This is fantastic news for anyone who is busy, and who isn't? We already have enough to think about.

Changing Our Eating Patterns

Our bodies are constantly adapting to changing conditions. Thus, when we start a new diet or exercise program, our body adapts to help us do what we want to do. It can take four to eight weeks to build up adaptations but just two weeks of de-training (not training) to lose them. Therefore, we may not want to spend a lot of time building up adaptations we aren't planning on maintaining. It might be better to pick training that we will be able to easily continue. This also may help explain why extreme diet and exercise approaches for weight loss and muscle growth usually don't provide long-lasting results. As soon as a person stops the special weight-loss diet or backs off from the rigorous weight-training program, the fat comes right back on (and perhaps then some) and the muscle declines. Thus, we want to find a lifestyle that we can comfortably and conveniently continue long term that provides us the adaptations and results we're looking for. Answering hunger with fastercise is a great fit for many of us.

Using fastercise to direct our bodies to burn fat and build muscle is quite simple. When the body uses hunger to signal that it is starting to burn muscle—and "asking" us what fuel we'd like to burn—we answer with fastercise. This signals our bodies to release and burn stored fuel instead of muscle and our hunger goes away. Crisis averted. We remain comfortable as our body burns stored fat. In essence, we are replacing some of our usual food intake with fastercise. There are a variety of strategies we can use to plan when we will fastercise instead of eat.

Most Americans have gotten in the habit of eating several times a day: breakfast, lunch, dinner, and several snacks between meals. There's increasing evidence that this high frequency of eating can have damaging consequences for our health, and I delve into that in chapters 2 and 3. The good news, though, is that fastercise can also help us stop eating so frequently throughout the day. For example, you can use fastercise to stop eating between-meal snacks. Or, if you want to make a more dramatic change, you use fastercise to help you cut back to two meals a day, or just one meal a day. Another approach is to limit yourself to eating only within a six-to-eight-hour window of time each day (for example, from noon to 8 p.m., or 8 a.m. to 4 p.m.). This eating pattern is sometimes called intermittent fasting, and it's a part of several popular diet programs. Overall, the goal is to spend more time not eating, but also to avoid spending time fighting hunger pangs, because that would mean you're burning valuable muscle.

With practice, we can manage our hunger to more closely fit our schedule. Generally, our hunger depends on how much and how long ago we ate. If we find that our hunger is coming on too strong too early, or not early enough, we can consider changing the size or timing of our meals. If we find that we are having a hard time pushing off hunger with fastercise, then we can eat a little sooner. If we find that we aren't that hungry and we can easily push off hunger and burn more stored fat, then we can delay eating. Even if eating is your favorite activity and you don't want to skip any meal, you can still pay attention to signals and be sure to stop eating when your hunger goes away. That way, you'll get hungry again sooner, which is great, because studies show that it's more enjoyable to eat when we're hungry than when we're full.[1]

We can easily settle into a very comfortable and convenient balance between eating and fastercise, within a matter of days. When we do, we settle into a sustainable process of burning fat and building muscle. Every day that we make a little progress, we are setting the stage for making huge progress over time. How does that work? Well, imagine taking a twenty-inch step forward every morning and a twenty-inch step backward every night. If you did that every day for a year, you wouldn't make any progress; you would be right where you started. On the other hand, if you take a one-inch step forward every morning but don't take *any* step backward at night, then within a year you'd have progressed thirty feet! That is huge in comparison. This is the kind of astonishing progress we can make when we answer hunger with fastercise day by day. From one day to the next, you may not notice the difference but the results can accumulate surprisingly quickly and even seem unreal (if no progress has been your past reality) because even small progress is real progress. And because fastercise is so comfortable and convenient, we can continue to progress for weeks, months, and years, getting stronger, faster, leaner, and healthier day by day, instead of getting weaker, slower, fatter, and more unhealthy day by day.

Basic Fastercise Strategy

OK, you're ready to try fastercise, but you may still be wondering, how does this work on a day-to-day basis? The answer is that it's a little different for everyone, depending on what your daily schedule is like. But here is a little basic strategy to consider.

Tightercise is something you can do easily and naturally wherever you are. You can do it at your desk at work, or even during a meeting, without anyone noticing. Simply tighten the muscles you want to build, even without changing position. Or you may decide to tightercise by doing a big stretch like a morning stretch. This appears perfectly natural, because it's normal for anyone to stretch from time to time during the day. Our bodies stretch automatically from time to time during the day to mobilize energy stores, but we can do it on purpose to eliminate hunger, burn fat, and stimulate muscle growth.

"I liked the idea of using fastercise strategically in the workplace. I am an ER doctor and often there are tempting foods in the ER for hurried fast-paced workers that satisfy the cravings that come up in stressful situations (which tends to be the entire shift). I have fastercised while at my desk, and it did help abate food cravings. The other health care providers near me didn't even notice or know that I was effectively performing health management by burning calories, fending off cravings, and feeling more in control within five to ten minutes of fastercise. I continue to use fastercise today." —JOHN W.

When you use tightercise in this way, you may find that you are able to make it comfortably to lunchtime without eating breakfast. And that's fine because it's not as though you're starving yourself; you are simply signaling your body to burn stored fuel for breakfast instead of food fuel. Simple. Instead of eating breakfast, you're "snacking on storage," fat you've been carrying around from a pizza or hamburger you ate months, or even years ago.

At some point, you will decide to eat again. This may be when it becomes too distracting or inconvenient to fend off hunger with tightercise (because your appetite is coming back sooner and sooner). This may be around lunchtime. You should tightercise once more right before you eat. (I explain why this is so in chapter 6.) After the meal, as your day goes along and you start to feel hungry, you can perform another bout of tightercise. Again, it's a good sign when you find yourself catching a deep breath (typically, after less than a minute of tightercise). Often, your hunger will go away by the time you're finished with a bout of tightercise. This little bit of fastercise is enough to signal the body to mobilize stored fat and glycogen and to build muscle. The body complies and your hunger goes away, likely for 30 to 120 minutes or longer.

By the way, fastercise also releases epinephrine and other hormones that promote energy and mental focus and prepare us to handle stress. In fact, epinephrine is what caffeine mimics. Thus, fastercise releases built-in

Table 1.2. Simple Laws of Fastercising

Upon awakening	Shivercise until you can catch a deep breath; consider tightercise and dietary supplements.
Whenever you start feeling hungry	Fastercise with or without drinking water to push off hunger (may repeat one or more times), or fastercise and then eat.
Right before you eat	Fastercise to signal the body it doesn't need to store fat.
While eating	Relax, de-stress, and enjoy yourself.
When your appetite is just satisfied	Eat just enough to satisfy hunger and then stop eating.
When preparing for bedtime	Consider taking dietary supplements, going to bed earlier and getting up earlier, and using fewer covers (to prompt the body to burn more fat to stay warm).

fuel and epinephrine while coffee, cream, and sugar supply food fuel and caffeine instead. Thus, fastercise is like a built-in cup of coffee, only better, since it also stimulates muscle and mitochondrial growth and so much more. Why spend so much time and money on fattening drinks when we can easily fastercise instead?

There's lots more to say about fastercise strategy, and I offer many more tips and ideas in chapter 6. One question many people have is whether they need to go on a low-carb diet if they plan to start fastercising. I tell them no, fastercise can be effective with any healthy eating plan. I also believe it's possible to adapt fastercise to either a high-carb or low-carb diet, and I discuss these diets in detail in chapter 9. One advantage of a high-carb diet is that most people already eat a high-carb diet. So following a high-carb diet makes it easier to share meals with friends and family. One disadvantage of a high-carb diet is that when hunger starts to come back after being pushed off with a bout of fastercise, it tends to come back more aggressively. On the other hand, one advantage of a low-carb diet is that it can help us burn more fat while perhaps building more muscle. One disadvantage of a low-carb diet is that we need to be sure to get enough water and electrolytes to feel our best.

At the end of the day, it's always a good idea to get plenty of sleep to allow your body to rest and repair. Sleepiness is also a signal the body sends that we can heed to our benefit. I have found with regular fastercise that I need less sleep and feel more rested. Each day you can arise and repeat your fastercise strategy, and you'll find it becomes more and more natural over time. If you like, it's perfectly fine to take a day off from fastercise and your usual eating pattern every so often (once a week, for example). The beauty of fastercise is that it can easily provide excellent results, which makes it extremely sustainable. Not only is this an approach that almost anyone can follow to achieve extraordinary fitness, but it's also an approach that people can easily continue for the long term to maintain incredible health benefits and vitality.

CHAPTER 2

It's All About Energy

When you ride a bicycle to the top of a hill, you store up potential energy that you can use to pick up speed without even pedaling as you ride back downhill. In a similar way, plants store chemical energy that they gain through the process of photosynthesis, a complex set of reactions powered by energy from sunlight. When we eat and digest fruits and vegetables, we release that stored chemical energy by breaking down the biomolecules in those plants in such a way that electrons are passed "downhill" from one type of molecule to another and eventually to oxygen. The passing of electrons from fuel to oxygen is also the mechanism by which campfires generate heat. Campfires take in oxygen (O_2), release energy, and produce carbon dioxide (CO_2) and water (H_2O) as by-products of the process. Just as campfire logs go up in a smoke of CO_2 and H_2O, when we metabolize fat (high-energy storage biomolecule), about 84 percent of it leaves our body as CO_2 and about 16 percent leaves as H_2O. Thus, our bodies do *burn* fat, carbohydrates, and protein to produce energy—they just don't produce any flames!

In the big picture, plants produce what our bodies use, and our bodies produce what plants use. How convenient. Plants take in carbon dioxide, water, and energy from the sun and transform them into complex biomolecules and O_2. We breathe oxygen into our lungs, and from there it travels in the bloodstream to cells throughout the body. As we eat plants (or animals that have eaten plants), our bodies break down the biomolecules, releasing their stored chemical energy. In this process, electrons are captured and eventually combined with hydrogen atoms and oxygen to form water. The carbon atoms liberated are also combined with oxygen to form CO_2. Our blood collects this CO_2 and carries it to our lungs. We breathe out water vapor and CO_2. Perfect.

When food is not available, our bodies can run on stored energy instead. Our bodies storing energy for later use is like charging a cell phone. Electricity drives chemical reactions in the phone's internal battery that build up high-energy storage molecules. Then, when the phone is disconnected from the power source, the chemical reactions begin to run backward and deliver the stored energy back as electricity. Our bodies build up stored power in the form of high-energy storage biomolecules: glycogen and fat. When food energy from our latest meal runs out, the chemical reactions that formed glycogen and fat can essentially run in reverse and deliver the stored energy back as chemical energy. One big difference between these two examples of energy storage is that when we leave our cell phone charging it will eventually fill up and stop accepting a charge, but if we keep charging our bodies with food, they can become bigger and bigger as they continue to store fuel.

Our bodies regulate whether we are in energy-storing mode (anabolism) or energy-consuming mode (catabolism). One way that chemical reactions in the body are regulated is by simple chemical equilibrium between reactants and products. Chemical reactions turn reactants into products. Frequently, reactions can also run backward at the same time to turn those products back into the reactants. The more reactants there are, the faster the forward version tends to go; the more products there are, the faster the backward version tends to run. When the speed of the forward and backward versions of the reaction equalizes, it is said that the reaction has reached equilibrium. In addition, the body can secrete hormones that have a great impact on the progress of chemical reactions, and I explain more about those hormones in chapter 3. For example, the thyroid hormone T3 increases the metabolic rate (the speed at which the chemical reactions in the body take place) by increasing the expression of DNA in the nuclei of the body's cells.

Changing the body's direction from anabolism to catabolism can take time. It's like driving a car. The farther you drive from home, the longer it will take you to get back home. So if you've been driving your body toward energy storage (anabolism) and obesity, it will take some time to change directions, gear up into catabolism, and get back to normal weight.

Energy Transfer in the Body

Our bodies can access the energy stored in a variety of fuels, and different parts of our bodies rely on different fuels to meet their energy needs. For example, resting skeletal muscles derive 85 percent of their energy from fats, but for bursts of energy, muscles burns glucose from food and glycogen. They can also burn ketones, which arc important breakdown products of fats. On the other hand, the heart muscle also relies heavily on fatty acids for fuel when the body is at rest, but its primary fuel during exercise is lactate, which is a breakdown product of glucose that is produced and released into the bloodstream by exercising skeletal muscle.[1] This is handy because exercising skeletal muscles can quickly obtain energy from glucose by breaking it down to lactate, and the heart is able to thrive on that lactate, which still contains the majority of the energy from the original glucose. Red and white blood cells and the retina and cornea of the eyes use primarily glucose for fuel. Intestinal cells absorb glucose for use by other tissues, but they themselves rely primarily on an amino acid called glutamine as their energy source. The kidney also can use glutamine for fuel. Interestingly, although one of the liver's important functions is to provide glucose and fat to other tissues for use as fuel, the liver relies primarily on breakdown products of amino acids to supply its own energy needs.[2]

As we digest carbohydrates, proteins, and fats in food, they are broken down into smaller molecules that are absorbed into the bloodstream by the intestines. The nutrient-rich blood from the intestines goes directly to the liver, which enables the liver to evaluate the levels of incoming nutrients and help manage and regulate the use of those nutrients. The body uses some of the nutrients to provide energy right away, but it can also build carbohydrate, protein, and fat energy stores for later use. The storage form of carbohydrate is glycogen, the storage form of protein is muscle, and the storage form of fat is fat. The fuel storage in the body for a normally proportioned man is about 2 percent glycogen, 31 percent mobilizable muscle, and 67 percent fat.[3]

Glucose and Glycogen

Carbohydrates, as the name implies, are comprised of carbon and water. The body breaks down complex carbohydrates in food into glucose, a small carbohydrate molecule containing six carbon atoms. Two-thirds of the glucose absorbed by the intestines from food is converted by the liver into glycogen, fat, cholesterol, and other molecules. The other third of the glucose passes right through the liver for use as fuel by the body.

I like to describe a glycogen molecule as a waterlogged tree made of glucose. Individual glucose molecules are linked together in a branching structure. Unlike fat, which repels water, glycogen absorbs water, and one pound of glycogen holds three or four pounds of water. When the body burns a pound of glycogen, those three or four pounds of water are released.

The human body can store about one pound of glycogen, and that pound can generate about 1,800 kcal of energy. Roughly 80 percent of the body's glycogen is stored in muscles to provide quick energy for muscle movement. About 20 percent is stored in the liver and used to maintain blood sugar levels (more about the importance of blood sugar in chapter 3).

When a person does intense exercise, their entire glycogen store can be consumed in as little as twenty minutes. Less intense exercise can use up the stores within ninety minutes, and simply going without food can deplete glycogen storage within about twenty-four hours. It can take one to two days to fully replenish fully depleted glycogen stores.

When we break down our glycogen stores for fuel, our bodies can recycle some of the breakdown products back into glycogen. Our bodies also replenish our glycogen stores using glucose derived from the carbohydrates and proteins we eat.

Fat

When our food provides more energy than we can immediately use or store as glycogen, excess carbohydrates and proteins can be broken down and stored as fat. By far, we store most of our energy in the form of fat. Although we can only store enough glycogen to last about one day, stored fat enables most people to live for many weeks without food. Our bodies are well equipped to rely on fat as an energy source, even though it may not seem like it sometimes. When people lose weight during a period of starvation, their bodies will burn fat and try to preserve muscle.

How ATP Is Produced

When people travel to other countries, they often find it handy to exchange some money for the local currency. They lose a little value in time and processing fees, but having local currency can save them time and money in their local transactions. Similarly, it's handy for the body to exchange the chemical energy stored in food for the body's local energy currency, which is molecules of ATP (adenosine triphosphate). ATP is found in *all* forms of life. When our cells use chemical energy from food to make ATP, 66 percent of that energy is lost as heat. When the chemical energy that is stored in ATP powers construction and movement in our bodies, 66 percent of that energy is also lost as heat. The bigger the energy loss in a chemical reaction, the more likely it is to take place. You can compare this to a ball rolling down a hill. If the slope is very gradual, maybe the ball will roll, maybe it won't. The steeper the hill, the more likely the ball is to roll. If the slope is so steep that the ball can lose 66 percent of its altitude very quickly, the ball will almost certainly roll. All this energy loss during the body's chemical reactions is a good thing because the released energy helps keep us warm. Thus, the body generates heat when it stores energy from food as well as when it uses energy from food and from storage. No wonder our body temperatures are about 25°F warmer than standard room temperature!

Our understanding of how the body works changes over time. Information about metabolism is widely available in reference books, journals, and on the internet, but it is a mixture of old understanding that has been passed along for decades and new understanding based on recent evidence. The explanations in this book provide a different perspective than found in some textbooks. My current understanding is based on a wide range of old and new information I have sifted through. I have no doubt that despite my best efforts some of what I write will be found to be incorrect in the future, but even so, I hope that what I write will be of benefit to many who read this book.

As I explained in chapter 1, humans need to be able to access energy quickly enough to escape dangerous situations. We also need to be able to glean as much energy from our food as possible to minimize our chances of starving to death. In other words, sometimes we need speed, sometimes we need endurance. It's fascinating to see how our bodies are set up to meet both of these needs perfectly. In a sequence of two metabolic processes, our bodies break down food and transfer the released energy into ATP. The first process breaks down food to a degree, generates some ATP, and then passes what's left of the food molecules to the second process. The second process breaks down the food even more and generates more ATP. The first process takes place in the cytosol of cells, so I call it the *cytosolic process*. *Cytosol* is the watery part of the cell that surrounds the nucleus and other organized structures. The second process takes place in mitochondria, which are organized structures suspended in the cytosol, and thus I call it the *mitochondrial process*. The cytosolic and mitochondrial processes collectively add up to *cellular respiration*.

Mitochondria are often referred to as the powerhouses of the cell. One cell can contain thousands of mitochondria, and mitochondria are found in every multicellular organism! That's right, plants, bugs, worms, frogs, birds, squirrels, and us. This suggests how critical they are for capturing energy from food. At the start of this chapter, I described how the body combines food with oxygen in order to liberate energy and produce CO_2 and H_2O. This process occurs in the mitochondria. In the mitochondria, food molecules are "burned" by combining them with O_2 through chemical reactions to generate CO_2, H_2O, and ATP. The fact that we can die within several *minutes* without oxygen demonstrates how critical this process is for our survival. We can't endure long without it. Around 90 percent of the ATP in the human body is made in the mitochondria. Our bodies constantly generate and use ATP. In fact, at rest, we generate and use up about 75 percent of our entire body weight of ATP every twenty-four hours. Even more astonishing is that with physical exertion like running we can generate and use more than our entire body weight of ATP in just three hours.[4] As long as we have oxygen to breathe (and water to drink), we can live for weeks without food because our bodies have this remarkable capacity to burn stored fuel to make ATP.

The relationship between the cytosolic process and the mitochondrial process is where things get really interesting. The cytosolic process is built for speed, while the mitochondrial process is ideal for endurance. The cytosolic process can generate ATP one hundred times faster than the mitochondrial process. This is crucial for fast energy generation (for escaping dangerous situations). The cytosolic process can sustain all-out exertion for about one minute, perhaps less, depending on an individual's physical condition. After that, the exertion intensity automatically decreases because the body has safety mechanisms to prevent tissue damage that would otherwise occur.

On the other hand, the mitochondrial process is about fifteen times more thorough than the cytosolic process at transferring food energy into ATP. This process is critical to extracting enough energy from food to provide physical endurance and to avoid starvation. Without the mitochondrial process, you'd have to eat about sixteen times more food than you do now to reap the same amount of energy.

Both the cytosolic and mitochondrial processes happen continuously and automatically inside the cells of your body. And here's a key point: The cytosolic process doesn't have to wait for the mitochondrial process. In times of need, the cytosolic process can speed up to generate a lot of ATP very quickly to provide an immediate burst of energy. This results in a backlog of partially digested food molecules. (Remember, the cytosolic process doesn't break down food molecules completely.) Those partially digested molecules must wait on the slower, more thorough mitochondrial process.

Burning Carbohydrates

Let's look more closely at how this works, starting with the example of burning carbohydrates for energy. Glucose is made up of six carbons and the equivalent of six water molecules. In the cytosolic process, the six-carbon glucose molecule is converted into three-carbon molecules, first *pyruvate* and then *lactate*.

When we analyze the cytosolic process more closely, we find that it consists of two subprocesses. The first subprocess, which generates pyruvate and ATP, is called glycolysis. The second process, which converts pyruvate into lactate, is called fermentation. Fermentation happens in the cytosol but it also happens in the digestive tracts of all animals as well as in yeast and bacteria. (Fermentation produces the lactic acid that preserves food such as pickles, kimchi, and sauerkraut, as well as the alcohol in beer and wine.)

Both pyruvate and lactate can be transported into the mitochondria for further processing. When we need a quick boost of energy, the cytosolic process generates pyruvate and lactate faster than the mitochondria can process it. Pent-up pyruvate can be turned into lactate (which enables the cytosolic process to keep producing more ATP and pyruvate). Sometimes a buildup of lactate results. This lactate can be used by nearby muscle cells for fuel. Excess lactate can enter the bloodstream to be used by other tissues as well. The mitochondrial process breaks down the pyruvate and lactate to one-carbon fragments—CO_2, which is small enough to diffuse out of the cell and into the bloodstream, where it is carried to the lungs and exhaled.

Burning Proteins, Fats, and Ketones

Our bodies break down proteins in food into amino acids. The body uses some of those amino acids as fuel, and some as building blocks to make its own proteins. Those proteins serve many purposes and are used in almost every cellular process. Proteins are important as enzymes, as hormones, for immune response, and as building material for muscle and other tissues. As I explained in chapter 1, due to the great functional value of muscles and proteins, the body strives to preserve them rather than use them as fuel. Nevertheless, it is normal for the body to break down some muscle tissue into amino acids to supply fuel, especially when glucose levels are low. Amino acids are broken down through the cytosolic process and the mitochondrial process. Some amino acids can also be converted into glucose and ketones that can be released into the bloodstream and used by the brain and other tissues for fuel.

The fat we use for fuel is made up of triglycerides, which are large molecules composed of fatty acids and a carbohydrate called glycerol. Fatty acids are molecules that have long, medium, or short chains of carbons. These chains of carbon can either be saturated with hydrogen or unsaturated (hence, the terms *saturated* and *unsaturated fats*). Each triglyceride is composed of three fatty acids attached to a glycerol "backbone." When triglycerides are broken down for fuel, the fatty acids feed into the mitochondrial process and the glycerol feeds into the cytosolic process.

Ninety-eight percent of the fat in foods is in the form of triglycerides. When we eat triglycerides, enzymes in the intestines snip off the fatty acids, and those fatty acids are absorbed by cells of the small intestine. Those cells reassemble the fatty acids into triglycerides. They then "pack" the triglycerides into specialized proteins, which transport the triglycerides through the bloodstream. This is necessary because triglycerides are oils and blood is a watery medium, and as we all know, oil and water don't mix well. Along the walls of some small blood vessels, there are cells that have the capability to snip fatty acids off the triglycerides packed in the transport proteins. These fatty acids are transported into the cells of muscles, the heart, and fat tissues (as well as other tissues). Once inside the cells, the fatty acids are transported into the mitochondria, where they are broken down to generate ATP. In order to be transported into the mitochondria, a fatty acid must first attach to a molecule called carnitine. That's why carnitine is included in some supplements designed to support fat loss.

Ketones are four-carbon molecules that can quickly enter the cell and mitochondria through membrane transporters, without hormonal regulation.[5] Though our bodies make ketones from fats all the time, ketone levels tend to increase when we burn more fat for fuel. Fat burning can go up during exercise or periods of food deprivation. People who are on a low-carb diet or have uncontrolled type 1 diabetes also burn more fat. Ketones can cross the blood-brain barrier and provide fuel for the brain. Also, each ketone molecule provides about 28 percent more energy than each pyruvate molecule, which can be useful when we're exercising.[6]

The Krebs Cycle

It's important to understand that the utilization of carbohydrates, proteins, fats, and ketones as fuel takes place in a soupy mixture of molecules—or rather two soups, the cytosolic soup and the mitochondrial soup. For now, let's focus on the mitochondrial soup. The fuel molecules inside the mitochondrial soup enter into a sequence of chemical reactions or pathways referred to as the citric acid cycle (CAC), tricarboxylic acid cycle (TCA cycle), or Krebs cycle. Pioneering biochemist Hans Adolf Krebs first identified this pathway of chemical reactions in 1937. This pathway fuels the vast majority of ATP production. And not only is it the hub of energy production, it also plays a vital role in providing building blocks for the construction of proteins, DNA, RNA, hormones, vitamins, ATP, hemoglobin, myoglobin, and many other important compounds. Thus, the CAC provides our bodies with both energy and building materials. In the CAC pathway, chemical reactions convert one type of molecule into another into another into another until the last molecule in the cycle is converted back into the original molecule, forming a cycle, as shown in figure 2.1. This is similar to the way a bucket at the top of a waterwheel collects water (which represents food energy), releases it toward the bottom of the wheel, and returns to the top of the turning wheel to receive more water (food energy). These molecules are referred to as intermediates.

I like to compare the CAC to a roundabout, a circular intersection that allows traffic to flow in one direction around a central island. Vehicles flow into and out of the roundabout through the various streets that intersect with it. Such roundabouts can have one, two, or even more lanes of traffic. Carbohydrates, proteins, and fats can all be broken down to contribute to the formation of one or more of the intermediates in the CAC. Like vehicles entering a roundabout, these breakdown products flow into the CAC at various points of the cycle. Intermediates can also flow out of the CAC to be turned into glucose for the brain or to provide building materials.

Unlike a roundabout made of asphalt, our CAC roundabout is expandable and collapsible, which is extremely efficient. When a single-lane asphalt roundabout is flooded with traffic during an evacuation, for example, there will be a huge backup because only so many vehicles can fit through one lane of traffic. But the capacity of the CAC expands and contracts as

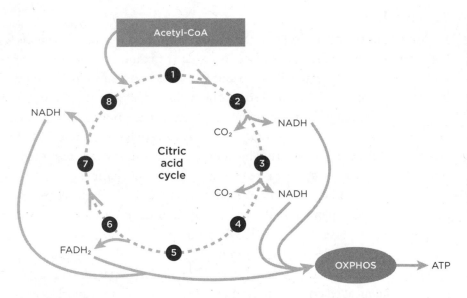

Figure 2.1. We can think of the citric acid cycle (CAC) as a cyclical series of chemical reactions that produces CO_2 and compounds called dinucleotides (NADH, $FADH_2$) as by-products. Those dinucleotides then feed into a process called oxidative phosphorylation (OXPHOS) that leads to production of ATP.

needed. When extra intermediates are added to the cycle, it's analogous to adding more lanes of traffic. When intermediate molecules leave the CAC to serve as building materials, that is like removing lanes of traffic. There are even words for this. *Anaplerosis* is the term for expanding the quantity of intermediates of the CAC, and *cataplerosis* is the term for reducing them.

Carbohydrates, proteins, fats, and ketones can be broken down into a molecule called acetyl-CoA, which is the fuel source of the CAC. Acetyl-CoA is the "traffic" that generates energy. It's easy to see that we can generate the most energy when both the fuel (traffic) and intermediates (lanes) are plentiful.

When you consume a lot of carbohydrates or protein and generate a lot of glucose, that glucose is converted to pyruvate, which is transported into the mitochondria. There, some of the pyruvate can be changed into acetyl-CoA, which can enter one lane of "traffic" by combining with an intermediate of the CAC called oxaloacetate. With one turn of the round-about oxaloacetate is re-formed. At the same time, some of the pyruvate can also be turned into extra oxaloacetate directly. That's like adding a second lane of traffic. This process can continue with as many lanes of traffic being added as needed to efficiently process food, generate energy, and produce building materials.

If you've eaten more carbohydrates or protein than you currently need for energy, your body may determine that you don't need to burn it all right now. Instead, the excess fuel (acetyl-CoA) can combine with oxaloacetate to form a compound called *citrate*. The citrate is transported out of the mitochondria. Once in the cytosol, that citrate is broken back down into oxaloacetate and acetyl-CoA. That acetyl-CoA can be used to make fat and cholesterol (this is how excess carbs and protein can be stored as fat in our bodies). The oxaloacetate can be converted into pyruvate, which can be transported back into the mitochondria.

Usually the body as a whole, with all of its tissues, is either in the state of storing energy or mobilizing and using stored energy. You'll recall from earlier in this chapter that *anabolism* is the term for the use of energy from food to build biomolecules for function and for energy storage. *Catabolism* is the term for the breakdown of those biomolecules to provide energy and recycle building materials to make needed biomolecules. Although many experts would disagree with me, I believe it is possible for muscle tissue to be in anabolism *at the same time* that fat tissue is in catabolism. I will present evidence for this statement later in this book.

Body composition is a term for the proportion of fat and fat-free mass in your body. The relative proportions of fat, muscle, bone, and water in our bodies help make up our body composition. Sometimes, when people start watching what they eat and exercising more, their body composition will change even though their weight stays the same. They will have more muscle and less fat, so obviously, they were able to gain muscle while losing fat. But for now, let's discuss anabolism and catabolism on the scale of the body as a whole in terms of gaining or losing weight.

We can think of anabolism and catabolism as directions of travel. Picture yourself traveling on a road between the city named Leanness and the city named Obesity. When you are traveling in the direction of anabolism, you are heading toward Obesity. When you turn around and travel in the direction of catabolism, you are heading toward Leanness. For this analogy, let's say there's no parking or stopping allowed on the road because you must be either in anabolism or catabolism—if you stopped, that would mean you were dead.

Some people are able to maintain a normal weight by driving in a loop, spending some time in anabolism, turning around, and then spending

A Critical Separation

It's fascinating to see how the processes of anabolism and catabolism are separated in our cells. Carbohydrates, fats, proteins, and ketones are broken down to provide energy (catabolism), and this occurs mostly in the mitochondria. The process of building up the substances the body needs (anabolism) happens mostly in the cytosol. Anabolism and catabolism are physically separated by the mitochondrial membranes. If the anabolism and catabolism weren't separated, it would be an inefficient mess, like trying to drive your car by pressing the gas pedal and the brake at the same time.

some time in catabolism. This is like driving around the block—you are always in one state or the other, or transitioning between the two, and you stay in the vicinity where you started.

Hunger is a great indicator of whether you are in anabolism or catabolism. Your hunger going away when you eat a meal suggests you are in anabolism. As your body finishes processing a meal and you begin to get hungry, that indicates you are turning around and heading into catabolism. If you are driving a car, you turn the steering wheel to change the direction you are driving in. The switch in your body between anabolism and catabolism, though, is regulated by hormones, and that brings us back to learning how to send your body signals that help promote hormones that direct your body to burn stored fat.

The Body's Response to Hunger

It's critically important that the brain work correctly, reliably, and quickly. In order to do so, it needs a very reliable supply of fuel, preferentially a carbohydrate such as glucose and lactate (which the body generates from glucose) and oxygen. A great deal of how our bodies respond to hunger has to do with supplying our brains with fuel.

Sources of Brain Fuel

It may surprise you to learn that the brain has very little chemical energy storage capacity of its own. The brain relies almost exclusively on energy supplied by the bloodstream from other parts of the body, primarily in the form of glucose. Why would that be? One way to think of it is that the brain is so important that the entire body works to keep it supplied with fuel. Another reason is to minimize the intracranial space used on fuel storage in order to maximize the space available for neuronal connections. Also, neuronal connections are optimized for speed, and glucose is a faster fuel than fat is.

It is commonly said that the reason our brains rely on glucose instead of fat is that fatty acids do not cross the blood-brain barrier. However, one recent study has demonstrated that fatty acids can cross the blood-brain barrier quite readily.[7] Other researchers offer a different explanation for the brain's preferential use of glucose, pointing out that the mitochondria

Brain Size and Ketosis

How readily a person goes into ketosis depends on the size of their brain relative to their body. An infant's brain is much bigger in relation to the rest of its body than that of an adult. Consequently, children and especially infants go into ketosis much more readily than adults. Interestingly, bears do not go into ketosis even when hibernating, because their brains are so small in relation to the rest of their bodies that they can make enough glucose for their brains just from the glycerol released from the breakdown of their fat.

in the brain have very little capacity to process fatty acids. These researchers hypothesize that the three main reasons the brain uses glucose for energy instead of fat are that glucose generates energy faster, burning fat is more stressful (oxidative stress) on the brain, and burning glucose uses up less oxygen.[8]

The three main fuels that supply energy for the brain are glucose, lactate, and ketones. The brain relies heavily on glucose as long as it is readily available from food or glycogen stores. When we are at rest, the brain commandeers about 60 percent of the total glucose used by the body! And, when our food and glycogen supplies become depleted, our bodies begin breaking down muscle in order to make more glucose. It is common knowledge that in this situation, the brain also begins to rely more heavily on ketones for its energy. In addition, recent research has demonstrated that our brains also use lactate as a fuel, apparently to help us conserve glucose, and thereby protein. The brain can use lactate and ketones directly as fuel, or the lactate and ketones may be converted to glucose first; this conversion occurs in the liver and kidneys. Relying more heavily on these alternate fuels helps the body preserve its protein (muscle). In fact, death by starvation is not due to the body running out of fat. Death occurs when the body runs out of protein.

The Hunger Signal

When the glucose derived from a meal begins to run out, amino acids such as alanine and glutamine begin to build up in the bloodstream. This is an indicator that the body is breaking down protein, especially from muscle, in order to generate glucose. The buildup of alanine and glutamine in the blood appears to be what signals the breakdown of glycogen stores to generate glucose (a process called glycogenolysis) as well as the increased production of glucose from amino acids, lactate, and glycerol (a process called gluconeogenesis).[9] Thus, at the same time we start to get hungry as the energy from a meal runs out, these two amino acids are building up in the bloodstream. This is one reason I think of hunger as a signal that we're burning muscle.

The more that glycogen breakdown can provide glucose, the less the body will need to break down muscle to make glucose. By the same token, relying more heavily on lactate and ketones for brain fuel as glycogen becomes depleted also helps the body preserve protein.

Our brains can use lactate for fuel not only during times when food is scarce but also when we generate lactate by exercising. Earlier I mentioned how the cytosolic process in muscle cells can generate a lot of lactate that can enter the bloodstream. When this occurs, lactate levels can rise, and when lactate levels are high, the brain takes in more lactate from the

More About Gluconeogenesis

People often talk about gluconeogenesis as a process that occurs in the liver, but a significant amount occurs in the kidneys as well. Alanine and glutamine make up 50 percent of the amino acids used by the body to generate glucose. Alanine is converted to glucose in the liver, but glutamine is predominantly converted to glucose in the kidneys. Normally, 20–25 percent of total gluconeogenesis takes place in the kidneys, but that can increase to nearly 50 percent during starvation.

blood and uses more of it for fuel.[10] This may be part of how fastercise can eliminate hunger so quickly: The production of lactate by your muscles during fastercise satisfies the energy needs of the brain and other tissues, and hunger goes away.

When lactate levels are low (when a person is at rest) and the brain is active and using a lot of glucose quickly, the brain can also export extra lactate into the bloodstream, which can be used by the muscles for fuel. Thus, muscles and brain can release lactate into the bloodstream for use as fuel by muscles, the brain, and other tissues. Or the lactate can be converted back into glucose in the liver and kidneys, which is then released into circulation and used for fuel.[11]

Supplying Brain Energy During Fasting

The conversion of glucose into lactate and the recycling of lactate back into glucose are happening all the time and are responsible for a large percentage of the glucose used by the brain. The percentage increases the longer a person goes without food. One study showed that after twelve hours without food the percentage of glucose derived from lactate was 41 percent, after twenty hours it was 71 percent, and after forty hours it was 92 percent![12] This is another way the body preserves muscle. The more glycogen is depleted, the more the body must rely on making glucose from the recycling of lactate, the breakdown of muscle, and from ketones.

As fasting continues, the brain begins to rely much more heavily on ketones for energy than on glucose. This has a tremendous sparing effect on protein (because, as noted previously, ketones are a breakdown product of fat). As glycogen runs out and fasting continues, the body continues to burn fat. Eventually, the breakdown of fat begins providing more energy than the body is using. When that happens, the mitochondrial process begins to slow down, and the body begins to shunt excess acetyl-CoA primarily to the liver, where it is converted by mitochondria into ketones. When the level of ketones builds up in the body (a state known as ketosis) and the brain increases its use of ketones for fuel, the breakdown of muscle decreases significantly (as shown in figure 2.2). Nevertheless, even in ketosis, glucose still makes up 30 percent of the fuel used by the brain (that glucose comes from recycled lactate, pyruvate, glycerol, amino acids, and ketones, in that order.)[13] Thus, during a prolonged period without food,

Getting the Big Picture

Glucose from food feeds into a pool of carbohydrates (glucose and lactate) that the body maintains to provide quick energy to the brain. When glucose from food starts to run out, our body starts breaking down glycogen and muscle to continue feeding the pool. I like to use the term *circulating pool* because the body continuously makes lactate from glucose and remakes glucose from lactate.

Figure 2.2 shows that as the fuel from a meal runs out, the body begins to use all three fuel stores (glycogen, muscle, and fat/ketones). The body relies heavily on glycogen at first, while gluconeogenesis from muscle and other sources has a chance to build up along with the breakdown of fat (lipolysis). Hunger begins when the body transitions from relying on food as a fuel source to relying on body-stored fuel. By transitioning from food to glycogen as a main supplier of energy, the body spares muscles and other proteins.

Notice also how the body transitions from relying mostly on glycogen to relying mostly on fat as the fuel source. When the body transitions to fat, less muscle is used for fuel.

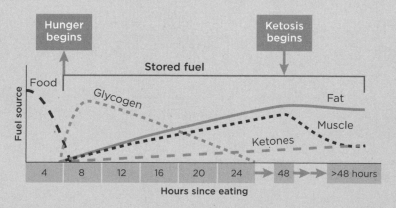

Figure 2.2. After a meal, the body passes through several transitions in fuel sources.

up to 70 percent of the brain's energy comes from ketones and 30 percent comes from glucose. Of that 30 percent, up to 90 percent of the glucose comes from recycled lactate. That leaves only 10 percent of 30 percent, or 3 percent, coming from other sources such as glycerol and amino acids. It's easy to see how this greatly reduces the body's need to break down muscle to provide glucose.

Although the body can turn carbohydrates and proteins into fat, it can't turn fat into glucose to help fuel the brain. The glycerol component of fat can be used to make glucose, but only a tiny fraction of the energy stored in fat is stored in the glycerol portion. The vast majority of energy stored in fat is stored in the fatty acids. Fatty acids are broken down by the mitochondrial process into two-carbon fragments (acetyl-CoA). For every molecule of acetyl-CoA that enters the CAC, two carbon atoms are released from the CAC in the form of CO_2. Therefore, no carbons are left to provide a net increase in glucose. However, *energy* generated during the breakdown of fat can be used to power the process that makes glucose from amino acids and lactate. And that glucose can be used as fuel by the brain and other tissues, or it can be used to replenish glycogen. In this way, our bodies can significantly replenish our glycogen stores even when we are on a low-carb diet.[14] I discuss this aspect of a low-carb diet, and how fastercise can enhance results of a low-carb diet, in chapter 9.

CHAPTER 3

Hormones
Send Key Signals

Timing is crucial in so many aspects of our lives. For example, you wouldn't drive across town to visit a friend who you knew was away on vacation. It would be a waste of your time. In a similar way, sending your body a signal at a time when it's not able to respond to that signal can also yield very disappointing results. This is one of the basic principles of fastercise: timing your efforts correctly to send a signal when the body can best respond. And since hormones are key signaling molecules in our bodies, understanding what hormones do and how they interact can help you get the timing right to maximize your fat loss and health gain efforts with fastercise.

Hormones transmit signals in the body like a key sends a signal to a lock. Whether it's a metal key that we insert and turn in a doorknob or a plastic hotel room key card that we insert into the door's electronic lock, the signal is the same: "This person has the key to unlock and open this door." Without the key, there's no admittance. Hormones are like keys that fit into locking mechanisms we call hormone receptors. Just as a locking mechanism in a door lock changes shape or configuration when activated by the key, so, too, do hormone receptors change shape or undergo some kind of change when exposed to the hormones that they fit. Specific receptors respond to specific hormones just as specific door locks respond to specific keys.

Hormones help regulate key bodily functions, largely determining what our bodies do under various conditions and circumstances. Hormones are produced in glands and are transported by the bloodstream to specific tissues. Hormones direct the function of the body a little like traffic police direct traffic. When we're driving a certain direction we may come upon a traffic policeman who instructs us to stop, continue, or even take a detour. We may want to go in one direction but we are stopped and directed to go in another. That interruption instantly changes our route. In a similar way, we can be rolling along on the road to Leanness and we may come upon a hormone that changes our course, or even turns us around and points us to Obesity.

Optimal bodily function often involves proper balance. It is normal for levels of many hormones in the body to rise and fall to help the body maintain the right balance. To maximize our ability to work with our hormones, we need to understand that both hormone *levels* and hormone *sensitivity* are important. A rising hormone level sends a signal to the body, but how well the body responds to that signal depends on its current sensitivity to that hormone. I like to make an analogy between the body's sensitivity to hormones and the way the human eye works. When exposed to bright sunlight, the photoreceptors in our eyes become less sensitive to light so they don't get overly saturated. This enables us to see in bright sunlight. But when we move from bright sunlight directly into a dimly lit room, it can be hard to see at first. That's because our eyes need some time to rebuild their sensitivity (by restoring visual pigments that were previously broken down by the bright light).[1] It can take twenty to thirty minutes for our eyes to adjust completely. And when the eye's sensitivity is fully restored, some of the photoreceptors become so sensitive that they can react to a single photon of light!

Prolonged elevation of a hormone's level can reduce the body's sensitivity to that hormone. That's why hormones communicate best when their levels go up and down. It's good for their levels to increase in order to send their signal, and it's good for them to decline so that we can rebuild our sensitivity to that signal.

Our body's response to a hormone can depend on the situation our body is in, such as whether we have just eaten, whether the energy from a meal is just running out, or whether it's been days since our last meal. It's important for us to understand how hormones work in different situations so that we can use them to direct our bodies to our desired destination and avoid unwanted roadblocks. In this chapter, I focus on the hormones that play the most important roles in energy metabolism in the body. One of those hormones, insulin, directs the body toward anabolism—toward storing energy. Two hormones, ghrelin and leptin, play key roles in arousing and dousing the sensation of hunger. Four other hormones—epinephrine, growth hormone, cortisol, and glucagon—counterregulate the action of insulin and direct the body to access stored energy through catabolism.

Insulin

Insulin is one of the key hormones that determines whether the body is headed toward Leanness or Obesity. When we eat, the pancreas secretes insulin and insulin levels rise, which signals our bodies to go into anabolism. Thus, rising insulin levels signal our bodies to build glycogen and fat

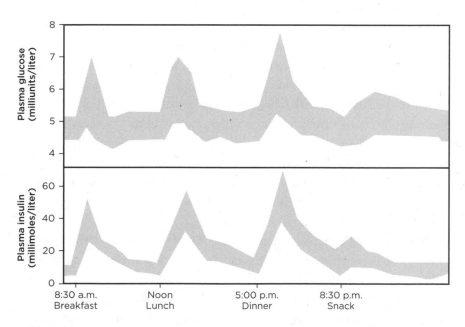

Figure 3.1. Although individual values for insulin and blood sugar (plasma glucose) levels vary among individuals, the overall trend is for insulin and blood sugar levels to rise when we eat a meal and decline as the food energy from the meal runs out. Adapted from Michael Feher and Clifford Bailey, "Reclassifying Insulins," *British Journal of Diabetes and Vascular Disease* 4, no. 1 (2004): 39–42.

and preserve muscle. In his compelling book *The Obesity Code*,[2] nephrologist Jason Fung points out that weight gain or loss is more determined by insulin levels than caloric intake. In one study, patients with type 2 diabetes (in this type of diabetes, the pancreas still makes insulin, but the body is insulin resistant) were given dosages of insulin to normalize their abnormally high blood sugar levels.[3] At the same time, their caloric intake was reduced by three hundred calories per day. As it turns out, the patients gained weight. Yes, their caloric intake went down and their weight went up. How is that possible? This study demonstrates that weight gain or loss is not just a matter of how many calories are available to the body, but what our bodies do with those calories. What we do with those calories is largely determined by hormones. Increasing the insulin levels of the study patients directed their bodies to store more calories (in the form of weight) even though fewer calories were available. This concept is important, because shivercise and tightercise are focused more on affecting hormone signals than they are on burning calories, as I explain in chapter 7.

Many people face confounding roadblocks in their weight-loss efforts. They want their bodies to do one thing, but their hormones are directing their bodies to do something else. In a study that investigated which test subjects had a harder time losing weight and a greater tendency to regain it, the researchers found that about 75 percent of the time, those who had high insulin levels (and associated inflammation) had the most difficulty losing weight.[4] In other words, it appears that when people have a hard time losing weight, it's often related to signals or directions their bodies are receiving from hormones such as insulin. This also suggests that higher insulin levels result in a higher *weight set point*, which is the weight range your body is currently geared to return to.

Although we can fill volumes with information about the topic of insulin, I believe that there is still much more that we don't know about this hormone. We do know it's best for the insulin supply to be "turned on" at some times and "turned off" at other times. Rising insulin helps us store extra food energy for later use. Dropping insulin helps us use up our stores so that we'll have room to store food energy from the next meal and maintain a healthy level of reserve overall.

We can develop serious health problems when we drive up levels of insulin for too long a period of time. In *The Obesity Code*, Dr. Fung draws a parallel between overexposure to insulin and overuse of antibiotics. When bacteria are exposed to antibiotics, they can become resistant to those antibiotics. When the body is exposed to too much insulin for too long, the body can become insulin resistant. When that happens, the body has to produce more insulin to signal the body to store glycogen, proteins, and fat. This can lead to even more resistance, which makes it harder and harder to get the signal through. If glycogen stores and fat cells are already full, they are resistant to storing more. But it's important to avoid allowing glucose levels in the bloodstream to remain high for too long; otherwise, the body will produce more insulin in an effort to clear excess glucose from the bloodstream. This can lead to more insulin resistance, as well as inflammation and disease. In this situation the body is being pushed too hard and too far in only one direction.

The good news is that insulin suppresses muscle breakdown and permits muscle growth as long as there is an adequate supply of amino acids (derived from protein in foods).[5] But when we spend too much time with high insulin levels, we tend to store more and more fat. As we develop insulin resistance, insulin's signal does not get through as effectively, muscle breakdown is not suppressed as much, and we can experience muscle wasting.[6] In the insulin-resistance scenario, a person can gain fat and lose muscle. This helps explain how it's possible to develop a large abdomen and skinny legs at the same time. Type 1 diabetes is a condition in which the pancreas secretes little or no insulin. People with type 1 diabetes experience muscle wasting when their insulin is too low, and even people who don't have type 1 diabetes can experience muscle wasting due to insulin resistance (high insulin levels). We get the most benefit from insulin when its levels fluctuate to promote both signal and sensitivity.

Another important aspect of insulin is that it can contribute significantly to fluid retention. The presence of insulin causes the kidneys to retain sodium, which reduces the amount of water eliminated.[7] It turns out that this effect of insulin persists even when people develop insulin resistance. Many people with obesity experience both insulin resistance and hypertension. It's interesting that lowering insulin levels (by not eating) can greatly reduce both fluid retention and high blood pressure. (Remember, three to four pounds of water are stored with each pound of glycogen.) When we eat, insulin rises to signal the body to store glycogen;

water is required for glycogen formation, so it make sense that insulin signals the kidneys to excrete less water. When the energy from a meal begins to run out, insulin levels decrease, and the body begins breaking down glycogen for fuel. The water stored with the glycogen is released. With lower insulin levels the kidneys are free to excrete more water. This release of fluids also has an impact on blood pressure.

Epinephrine

Epinephrine plays an important role in preparing the body to fight or flee—situations in which the body needs fuel fast. Secretion of epinephrine by the adrenal glands inhibits insulin secretion and increases glucagon secretion by the pancreas. (Glucagon increases the breakdown of glycogen and fat, as explained in "Glucagon" on page 54.) Epinephrine also directly stimulates the breakdown of glycogen in the liver and muscles to liberate glucose. Epinephrine acts to provide fuel not only in an emergency but also as a meal runs out. When blood sugar starts to drop, epinephrine levels begin to rise. Thus, the body naturally uses epinephrine to fight hunger. This is a big reason why we can use fastercise to answer hunger, because fastercise increases epinephrine.

Epinephrine also provides other benefits. It is the chief mobilizer of fat stores. And epinephrine even helps us fight our battles on a microscopic level, because immune cells that fight infection have epinephrine receptors. Epinephrine helps with mental focus and learning. It simulates nerve growth factor, which is essential for learning.[8] In an emergency situation, mental focus is important, as is remembering afterward what we've learned from the experience. Epinephrine also fights symptoms of food deprivation, including preoccupation with food, difficulty concentrating, and feeling cold, hungry, weak, tired, dizzy, and sleepy.

Last but not least, epinephrine helps us start our day. In the early morning, our epinephrine and cortisol levels begin to rise. Simple stretching when we wake up is enough to increase the release of epinephrine and cortisol. Fastercise in the morning can jump-start this process.

Just as the body needs to maintain a healthy balance in insulin levels, it also needs to experience fluctuations in levels of epinephrine and cortisol. It's natural—and helpful—for epinephrine and cortisol levels to go up at times of physical, mental, or emotional stress in order to help us deal with challenging circumstances. After the stress has passed, those levels go back down. But if we spend an unhealthy amount of time under pressure (due to excessive stress), the prolonged exposure to high levels of epinephrine can lead to epinephrine resistance. You may have heard that stress can cause people to gain weight, and such weight gain is frequently attributed to elevated cortisol levels. However, epinephrine resistance may play a big role as well. In fact, in a review article published in the *Journal of Pediatric Endocrinology and Metabolism*, researchers suggested that epinephrine resistance has a more causal role in childhood obesity than insulin resistance.[9] Their research shows that obesity in children and adolescents is characterized by a chronic excess of epinephrine, which leads to epinephrine resistance. This resistance contributes to the active phase of fat accumulation by decreasing lipolysis or fat burning as well as increasing the storage of fat. Further, they found that epinephrine resistance is reduced with exercise.

Growth Hormone

Growth hormone helps the body burn fat and gain muscle. The pituitary gland secretes growth hormone (GH) during times of fasting and stress (such as intense exercise). Thus, it is likely that GH exerts most of its impact under such conditions. As our insulin level decreases, our GH level rises, stimulating the breakdown of fat and the release of fatty acids. Fatty acids inhibit the insulin-dependent use of glucose. This spares the glucose for use by the brain and other tissues, which in turn spares muscle.[10] During fasting, GH spares whole-body protein (including muscle, enzymes, and structural and transport proteins).[11] Thus, in a sense, as the energy from a meal runs out, GH takes over insulin's job of preserving muscle and protein. A lack of GH during fasting would lead to a 50 percent increase in protein loss. This effect is largely due to less fat breakdown, which means that fewer fatty acids are present to inhibit glucose usage and thereby spare muscle.[12] Thus, when we generate less GH, we tend to lose muscle and gain visceral fat. This is another mechanism that explains how it is possible for a person to simultaneously develop skinnier limbs and a fatter trunk. A medical term used to describe this condition of low muscle mass and high fat mass is *sarcopenic* (low in muscle) *obesity*.

GH has numerous other health benefits. Since GH stimulates fat break-down it can promote the onset of ketosis, which is helpful for those on a ketogenic (low-carb) diet (more about ketogenic diets in chapter 9). GH not only prevents the breakdown of muscle, but it also promotes muscle building. In fact, increases in lean body mass due to GH appear to be due to enhanced muscle building more than decreased muscle breakdown.[13] GH boosts testosterone, energy, strength, and endurance. It can also contribute to healthy libido and sexual function. GH is known to reduce high blood pressure and may reduce early markers of atherosclerosis.[14]

The action of GH is also part of how exercise provides us with a more youthful appearance. Skin cells have GH receptors that contribute to increased collagen production.[15] This collagen strengthens the connective tissue in our skin, hair follicles, tendons, ligaments, cartilage, and bone. This can result in a significant tightening of loose skin, less hair loss, greater ease of movement, less susceptibility to injury, and better wound healing. People who are deficient in GH have thinner, saggier skin that returns to normal with GH treatment, while people who have excess GH can develop thick, rough, and oily skin.[16]

When the energy from a meal runs out, the stomach releases the hormone ghrelin into the bloodstream. This hormone increases our appetite and stimulates the release of GH, which helps us protect our muscle. In fact, *ghrelin* gets its name from the phrase *GH release*. A two-day fast can lead to a fivefold increase in GH.[17]

As fasting continues, the brain starts using ketones for fuel and its requirements for glucose decrease, so there is less need for muscle and protein breakdown, and hunger decreases significantly. This decline in the sensation of hunger when muscle breakdown decreases is another reason that I interpret hunger as a sign that the body is burning muscle. High-intensity exercise such as fastercise also increases GH, but it's interesting to note that we secrete most of our GH while we are asleep. This timing suggests that we do a lot of our fat burning and muscle building while sleeping! So it really behooves us to get plenty of sleep. Wouldn't it be interesting if we could actually burn more fat by sleeping for eight hours than we could by sleeping for seven hours and jogging for one?

Ghrelin and Leptin

Ghrelin is referred to as the hunger hormone because it is the only hormone known to increase the sensation of hunger. When the stomach is empty, it secretes ghrelin (as do other parts of the gastrointestinal tract). Ghrelin circulates to the brain to increase the feeling of hunger. The stomach filling up with food is one signal that decreases the secretion of ghrelin, resulting in hunger diminishing. Ghrelin also stimulates glucagon release, and glucagon stimulates the breakdown of glycogen to provide glucose for fuel.[18] The effects of ghrelin are a big part of why the body turns more toward glycogen and away from muscle for fuel as the food energy from a meal runs out. Ghrelin stimulates hunger especially for energy-dense foods while also stimulating the reduction of energy consumption by the body. This helps to explain why we "slow down" when we're hungry. An extensive body of research demonstrates that injecting subjects with ghrelin increases food intake and over time leads to weight gain and gaining fat. Thus, if our goal is to reduce fat, it makes sense for us to try to keep ghrelin levels down by not letting ourselves go around hungry. There is also a peak of ghrelin secretion in the early morning (about 2 a.m.) that coincides with the GH peak. This GH peak promotes cell regeneration as well as muscle building and fat loss.

It might surprise you to learn that obese people tend to have lower ghrelin levels than lean people. One reason for this is that obese people tend to have high leptin levels. Leptin is sometimes referred to as the satiety hormone because it reduces hunger and increases satiety. Leptin is made primarily by fat cells throughout the body, but some is made by other tissues as well. Obese people have proportionally more fat and therefore tend to have proportionally higher levels of leptin. This can lead to leptin resistance. Leptin inhibits ghrelin (which is part of the reason that leptin reduces hunger).[19] With leptin resistance, there is less inhibition of ghrelin, and thus less ghrelin is needed to generate hunger. Due to leptin resistance, obese people can tend to get hungry more easily (less ghrelin is needed), and because they are resistant to leptin, they may need to eat more food in order to feel satiated. Nevertheless, this situation can be turned around as an obese person begins to lose fat. The body will make less leptin and will be less resistant to leptin.

Ghrelin and Fertility

Keeping ghrelin levels and appetite down may increase fertility. Ghrelin can also be produced by the gonads, and there are ghrelin receptors in the gonads as well. Though further research is needed to elucidate the effects of ghrelin on the gonads, it may be that ghrelin is a key signal letting the body know whether energy supply is favorable for reproduction.[20] Ghrelin receptors are found in the thyroid, adrenal, hypothalamus, and other tissues as well. (Conditions of starvation aren't conducive to child-rearing and fast metabolism.) Obesity contributes to infertility in both men and women.

High-intensity exercise such as shivercise increases leptin sensitivity more than low-intensity exercise (which helps explain why people often feel satisfied with less food after an intense bout of exercise). In one study, leptin levels decreased with exercise and decreased more significantly with higher intensity exercise, which is consistent with improved leptin sensitivity.[21] As you can see, it's important to consider not only the effects of hormones on the body but also the body's degree of resistance to those hormones.

As we eat and our insulin levels rise, leptin levels are stimulated to increase, too. In turn, leptin increases insulin sensitivity and signals the body that energy supply is adequate, which leads to decreased appetite. Leptin also decreases the liver's conversion of protein to glucose. This is yet another point that supports my belief that hunger is a sign that we are burning muscle. Hunger can be quite uncomfortable. Discomfort is one of the body's primary ways of warning us that something unhealthy is happening. When we start to get hungry our body automatically starts mobilizing our glycogen, muscle, and fat stores for energy use. At such a time, we can fastercise to enhance the mobilization of glycogen and fat stores, decrease the breakdown of muscle, and cancel our hunger. This supports the idea that the sensation of hunger is associated more with the breakdown of muscle than it is with the breakdown of glycogen and fat stores.

Cortisol

Cortisol has innumerable effects on the body; here I focus on the effects most relevant to the topic of energy usage, appetite, and exercise. Cortisol is produced in the adrenal gland and is released especially at times of stress and when blood glucose levels are low. To help maintain our glucose levels, cortisol aids epinephrine and glucagon in breaking down glycogen to release glucose. At the same time, cortisol decreases the effect of insulin on the tissues. In a sense, cortisol creates resistance to insulin, and that's why it's considered one of the four hormones (epinephrine, growth hormone, glucagon, cortisol) that are counterregulatory to insulin. Cortisol also has fat-burning effects. But just as cortisol can counter the effects of insulin, insulin can counter the effects of cortisol. Thus, the good effects of cortisol are especially accessible when insulin levels are low, when we are hungry.

When the body runs out of glycogen, cortisol levels can rise to maintain glucose levels through the breakdown of muscles.[22] This can be accompanied by a discernible sensation of stress and tension (hangry), which can be relieved by eating. Animals that have had their adrenal glands removed and cannot make cortisol literally cannot survive significant stress. As with other hormones, cortisol's signals are most effective when they are cyclic (sometimes low, sometimes high). Indeed, cortisol secretion normally follows a distinctly circadian rhythm; cortisol levels are higher at certain times of the day and lower at other times of the day. As with other hormones, when levels of cortisol are too high for too long, resistance to its signals can develop, leading to poor health.

Persistently high levels of cortisol can contribute to insulin resistance and lead to high levels of insulin, which can counter cortisol's fat-burning effects. This effect is one of the reasons some of us tend to gain weight under stress. When we are exposed to persistently high levels of cortisol we can tend to gain fat in some areas, especially around the organs in our abdomens, while losing fat in our arms and legs and under our skin. Scientists aren't exactly sure why this occurs, but research shows that different types of fat deposited in different areas of our bodies (called depots) respond to cortisol differently. For example, a 2013 article in *Cell Metabolism* showed that a particular gene (LMO3) is increased by cortisol, is more abundant in visceral adipose tissue of obese individuals, and promotes fat production.[23] LMO3 could help explain the increased accumulation of visceral fat in obesity due to excess cortisol. At the same time, peripheral subcutaneous depots can become depleted.[24] Persistently elevated cortisol can also contribute to muscle wasting. So here is yet another possible explanation of the tendency of some individuals to gain weight in the abdomen while losing weight in the arms and legs. This symptom, by the way, is commonly seen in people who have Cushing's syndrome (due to prolonged exposure to cortisol).

All of the above demonstrates how timing is everything when it comes to hormones. The very same hormone at different times and circumstances can help us burn fat, or store it. Based on the discussion above, keeping cortisol levels down when insulin levels are up might help us lose fat. This concept supports the wisdom of the ancient practice in many cultures of making mealtimes low-stress and relaxing occasions. In today's world, stressful working lunches and dinners may be making it harder for some of us to lose fat. Additionally, avoiding unnecessary stress when we're not hungry is a great idea because having lower cortisol levels when we aren't burning fat will help our bodies be more sensitive to it when we want to burn fat (when we're hungry). It also makes sense to answer hunger with fastercise to increase cortisol at times when insulin levels are low.

Glucagon

Glucagon is produced by the pancreas and is considered the main catabolic hormone of the body. It stimulates the breakdown of glycogen to raise glucose levels. It can also increase glucose levels by increasing the production of glucose in the liver and kidney. Glucagon also promotes the breakdown of stored fat when insulin levels are low. Like the other catabolic hormones that mobilize energy stores, glucagon is increased with high-intensity exercise.[25]

In this chapter we clearly see how important timing is when it comes to the effectiveness of hormonal signals. Like a baseball player up to bat, if we don't swing at the right time, we can easily strike out instead of hitting a home run. The actions of many hormones *oppose* or counter-regulate one another. In addition, it's healthy for hormone levels to go up and down under different circumstances. On top of that, when levels of some hormones are rising, levels of others can be declining. The complexity can be mind-boggling. With all of that going on, how in the world can we time our actions so we can hit home runs? How can we put the most important hormones of energy metabolism to work for us rather than against us? As you read this chapter, I hope you noticed that the discussion throughout points to a single simple approach to channeling the action of these hormones in the direction we want to go: burning fat, gaining muscle, decreasing hunger, boosting metabolism, and improving health. That simple approach is to answer our hunger with fastercise. BAM! Home run.

CHAPTER 4

The Non-Fed State

To maximize our entertainment experience, the movie industry uses the finest camera and recording equipment to carefully craft the signals that our eyes and ears receive. Movie theater designers focus on how to provide an environment that will maximize our sensitivity to what we will see and hear. The dim lighting improves our sensitivity to the images on the theater screen, and the acoustically padded walls that dampen ambient noises improve our sensitivity to the soundtrack. And, of course, the smells and tastes of the concessions can be quite mood-enhancing as well. All of this can make a big difference in the success of the experience. Imagine, for example, trying to watch a movie outside on a busy street in the noonday sun. It would be very difficult to enjoy the movie! In a similar way, we want to control the environment inside our bodies in order to maximize the success of our experiences. We want to increase our bodies' sensitivity to hormones by keeping their levels low when we want them to be low. That way, our bodies can respond optimally when we send a signal that causes those hormones to rise.

In a way, the metabolic state that we enter when the energy from a meal runs out is like dimming the lights and dampening unwanted sounds in a movie theater. I refer to this metabolic state as the *non-fed state*. When we're in the non-fed state our insulin level drops and our sensitivity to insulin begins to rebuild (insulin resistance decreases). This sensitivity will help us build muscle when our insulin level rises again. We get hungry, and the level of hormones that are counterregulatory to insulin (such as epinephrine and growth hormone) rise. These counterregulatory hormones help mobilize stored energy for use as fuel. Thus, entering the non-fed state is an easy way to turn our hormones around so our bodies are ready to burn more fat.

Another term for the non-fed state is the *fasting state*, and there are many diet plans that are based on some type of fasting protocol. I prefer to use the term *non-fed* instead of fasting because the definition of *fasting* is "to abstain from all or some kinds of food or drink, for a time, especially as a religious observance." This definition can make it sound like fasting is not something that everyone does. However, none of us eat and drink continuously. Every day, there are times when we stop eating and drinking. Also, the non-fed state doesn't begin the minute we stop eating; it begins at the time when the energy from food we've eaten begins to run out and we start feeling hungry. Abundant scientific research shows that spending more time in the non-fed state has lots of health benefits. This is yet another beneficial aspect of fastercise, because fastercise helps us control our appetite, which makes it easier to change our eating habits and spend more time in the non-fed state.

Many of us are spending too much time in the fed state, basically eating our way to disease and even death. Spending less time in the fed state can help us turn that around. Increasing the duration of deliberate caloric abstinence has been beneficial in treating rheumatic diseases, chronic pain syndromes, hypertension, metabolic disorders, and most chronic degenerative and chronic inflammatory disorders.[1] Obesity is stressful on the body. Obesity alone increases oxidative stress, which can lead to inflammation and insulin resistance. And since insulin resistance leads to obesity, this can become a vicious cycle.[2] Thus, while eating can be beneficial for our health, spending some time not eating is also an important part of good health, recovery, and recuperation. Spending more time in the non-fed state has been shown to promote neuroendocrine function, detoxification, nerve growth, reduced mitochondrial oxidative stress, refurbishing and replacement of damaged or worn-out molecules and cells, and the general decrease of signals associated with aging.

Let's carry the movie theater analogy a little further. If entering the non-fed state is like dimming the lights and dampening the sounds in the theater, then fastercising is like the carefully crafted movie that sends just the right signals to maximize fat-burning. Both are important. We need to enter the non-fed state and feel the hunger signal, and at times we can answer that signal with fastercise. However, we don't want to fastercise continuously, because we don't want to put undue stress on our bodies or

"I am a sixty-five-year-old professional that has fought weight issues most of my adult life. I have tried many different types of diets. Would lose weight, but in the end I put it back on. About a year ago, I first met with Dr. Wilson. Learning about fastercise, I was able to easily incorporate it into my busy life. With just a few minutes a day I was able to lose my first stage weight. When I stopped my healthy eating habits for a season, I noticed that I didn't gain the weight back as I did in the past. This gave me the encouragement to pick back up on the fastercise routine and continue back on track. By studying Dr. Wilson's methodology I am learning how to have a two-way communication with my body along with the hope that easy life changes will make a permanent difference."   —RON M.

become resistant to the counterregulatory hormones (epinephrine, growth hormone, cortisol, and glucagon). That would be like cranking up a movie projector's brightness level so high that the screen turns white—you can't see the movie at all. Sometimes it's good to eat; sometimes it's good not to eat. Sometimes it's good to fastercise; sometimes it's good to rest.

Fastercising while hungry (in the non-fed state) is a good example of working *with* our hormones rather than *against* them. Fastercise and the non-fed state both promote fat burning. Fastercise and the non-fed state both increase levels of growth hormone, which preserves muscle. In fact, fastercise increases the levels of all the counterregulatory hormones that increase during the non-fed state. Thus, by answering hunger with fastercise we are promoting the burning of fat and the retention of muscle, which is exactly the direction we want to go. Perfect.

Timing Versus Quantity

Anytime we eat, our insulin level goes up, which directs the body to store energy. Notice I didn't mention anything about eating a certain *quantity* of calories. A small amount of carbohydrate or amino acids triggers an increase in insulin levels, as little as a swallow or two of soft drink, a bite or two of chicken or potato, or ½ cup of broccoli. And the more time we spend with elevated insulin levels, the more time we spend on the road to Obesity. In chapter 3, I described a study that demonstrated that insulin has a greater bearing on fat loss and gain than the number of calories ingested. In another study of normal-weight middle-aged adults, participants were given equivalent amounts of calories, but some of them consumed the calories over three meals a day and others ate all their calories in one meal a day. The participants who ate one meal a day lost significantly more fat mass, simply because they spent more time each day with lower levels of insulin, or in other words, more time in the non-fed state.[3] Cortisol levels also decreased in the one-meal-a-day group, suggesting that eating frequently and accumulating excess fat may actually be stressful on the body. Another interesting study focused on people with type 2 diabetes. The study participants were given intensive insulin therapy to control blood sugar levels, and they ate a reduced-calorie diet for six months.[4] At the study's end, they experienced an average increase of about 10 percent in their body weight! As an example, that would be a twenty-pound weight gain for someone who weighs two hundred pounds. This seems confounding, because the subjects ate *less* but *gained* weight. Why? Because the insulin therapy was directing their bodies to store more of the calories that they *did* eat.

Sometimes it can seem as though our bodies refuse to lose fat, no matter what we do. What might cause this? One common scapegoat is a slow metabolism. The word *metabolism* simply means the sum of all the chemical reactions in the body. So, if the chemical reactions in your body are happening at a slower rate than they do in an average person, your body won't burn up as many calories, and you could have trouble losing weight. Surprisingly, though, a study published in *Clinics in Endocrinology and Metabolism* found that obese people actually have a higher basal metabolic rate than lean people.[5] This suggests to me that when we're overweight, our bodies are probably trying their best to lose weight, but how can they succeed if we persist in keeping our insulin levels high by eating all the time? Maybe we need to redefine our concept of overeating. Most people would define it as consuming *too many* calories. I suggest that spending *too much time* in the fed state is another form of overeating (regardless of whether the number of calories consumed is higher, lower, or the same as normal).

The term *intermittent fasting* conveys the idea of spending more time in the non-fed state than is conventionally the case. This could take the form of eating two meals a day instead of three. That's not conventional today in the United States, but it may have been in centuries past. Who knows? Maybe our bodies are built to spend more time in the non-fed state than is customary these days. It could be that many of us are overfed in the sense that we are spending entirely too much time in the fed state.

Did your mother or father ever warn you, "Don't snack, you'll spoil your dinner"? Decades ago, eating three meals a day with no snacking in between was more the norm. Obesity was less prevalent in those days. A study done at the University of Ottawa in Canada confirms that going without food for a time does heighten our enjoyment of food when we do eat.[6] This makes sense. The non-fed state can build up our body's sensitivity to and appreciation for food so that we enjoy it more when we do eat. The main challenge with spending a lot of time in the non-fed state is the discomfort of being hungry as well as feeling a lack of energy. Fortunately, you can cancel your hunger pangs and feel more energetic by fastercising for a matter of seconds instead of eating. Eating increases insulin levels, but intense exercise like fastercise decreases insulin levels, as does spending time in the non-fed state.[7]

Calories In, Calories Out

One prevalent paradigm used to predict fat loss can be boiled down to four words: Calories in, calories out. This leads us to think of an *energy balance* that determines our weight. We take in calories by eating and we move calories out by being active. This paradigm holds that as we eat less and exercise more we will lose fat. Thermodynamically the paradigm makes perfect sense, but as I noted above, it doesn't seem to explain what many people experience: They eat less and exercise more yet they don't lose fat or they may even gain fat. If it seems as if our bodies are fighting to keep their energy stores, it's because they are.

Let's compare your body's energy balance to a weekly spending budget. You earn $100 a week and you spend $5 a week on a hobby and $95 a week on everything else. At this rate, your savings can't grow. Then, let's say your income goes down to $95 a week, but you also make a change in your spending habits. You decide to put $10 a week into your hobby, but only $80 a week into everything else. Now, even though you are earning a little less money overall and spending a little bit more on your hobby, you have $5 a week left over to save for a rainy day. You'll be increasing your savings every week. Thinking about your body's energy balance, what if you were to cut back on calories but eat more frequently during the day? The way your body would react to this change depends a lot on your hormones—they largely determine what your body does with the calories you eat.

Two different conditions that both qualify as "eating less" generate very different hormonal responses. One is eating less food (consuming fewer calories); the other is eating less often. In the example above, you decided to eat less by consuming fewer calories, but you also decided to eat more often, which is a way of "eating more." Similarly, spending more time exercising, increasing the intensity of your exercise, or exercising more often all qualify as "exercising more." However, different forms and timings of exercise generate very different hormonal responses. The different hormonal responses to various combinations of these conditions of eating and exercise can have a huge bearing not only on whether we lose fat, but perhaps more important, whether we keep it off. The concept of "Calories In, Calories Out" is thermodynamically correct and irrefutable only when it means "Calories In, Total Calories Out," but usually people think of it as "Calories In, Calories Out via Exercise." Your body's hormonal response to the timing and combination of your version of "eating less and exercising more" can affect how many calories your body invests into storing fat or into building muscle. In chapter 5, I propose a new paradigm that can help us understand and remember how the body uses hormones and other signals to meet its own priorities.

Resetting the Set Point

Humans (and other animals) demonstrate a tendency to maintain a set weight for long periods of time.[8] After periods of overfeeding or underfeeding that result in weight gain or loss, animals tend to settle back to their original weights once they return to their normal eating habits. As I mentioned in chapter 2, this is the weight set point. No one knows for sure what determines this certain weight that our bodies want to weigh, but researchers believe that the phenomenon is multifactorial. It can be influenced by genetic and epigenetic factors as well as environment, diet, and physical activity. For example, prolonged overfeeding of animals has been shown to increase their weight set points and this seems to occur in humans as well. In *The Obesity Code*, Dr. Fung stated that the set point is affected by insulin levels.[9] High insulin levels raise our set point, and lowering our insulin levels can lower our set point. Dr. Fung identified that we can lower our insulin levels by eating less often. The authors of "Physiology of Sport and Exercise," pointed out that set point may be reduced by increased physical activity.[10] Other research has suggested that high-intensity exercise may have a stronger effect in this regard than lower intensity exercise. (I discuss this in more detail in chapter 7.)

The Minnesota Starvation Experiment is a famous study conducted in the 1940s to provide information that would help millions recovering

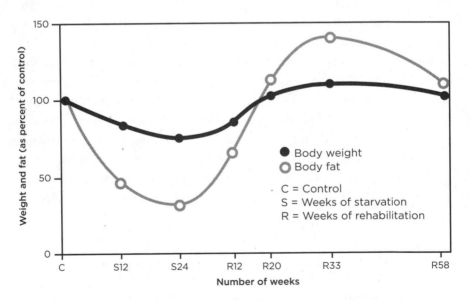

Figure 4.1. During the Minnesota Starvation Experiment, body weight and body-fat level decreased during the semistarvation phase. Upon refeeding, body weight and body-fat level rose significantly higher than levels preceding the semistarvation phase. Adapted from Ancel Keys et al., *The Biology of Human Starvation* (Minneapolis: University of Minnesota Press, 1950): 117.

from semistarvation due to World War II. Thirty-six men were selected
for the study, and researchers collected baseline information about the
men for twelve weeks. Then, for twenty-four weeks their caloric intake
was drastically reduced, resulting in an average weight loss of 25 percent
of starting weights, as shown in figure 4.1. After the semistarvation phase,
the men's normal calorie intake was restored for twelve weeks through
strictly controlled diets, followed by another eight weeks of unrestricted
but carefully documented diets. Thus, the whole study lasted over a year.
The men experienced many symptoms during the semistarvation phase,
some quite severe, including depression, hysteria, decreased libido, social
withdrawal, self-mutilation, decreased concentration, preoccupation with
food, and diminished metabolic rate.[11]

It's interesting to note that after the semistarvation phase, the men's
weight and body-fat level rose significantly above their prestudy levels for
a while. During the recovery phase the weight and fat levels continued to
settle back to levels similar to those at the beginning of the study, but still
remained slightly elevated. It appears that spending a prolonged period of
time in a state of semistarvation increased the men's fat set points. Many
people can relate to this experience of gaining back all their weight and
then some after dieting to lose weight.

Non-Fed, Less-Fed, More-Fed

At the beginning of this chapter, I defined the non-fed state, which occurs
when the energy supplied by a meal runs out and we eat nothing at all
for a time. There are other possible feeding conditions, too. The *fed state* is
a period of time during and following a regular meal. When we eat, our
hunger goes away, signaling our entrance into the fed state. When the
energy from that meal runs out several hours later, we get hungry, signaling
our return to the non-fed state. The fed state focuses our bodies on using
food for fuel and promotes the conservation and building of energy stores.

The fed state can be divided into three subsets: the *more-fed* state, the
normal-fed state, and the *less-fed* state. We're in the more-fed state when we
eat meals that have more than enough calories to cover our body's energy
consumption until our next meal. We're in the normal-fed state when we
eat meals sufficient to cover our energy usage until our next meal. We're
in the less-fed state when we eat meals that supply fewer calories than

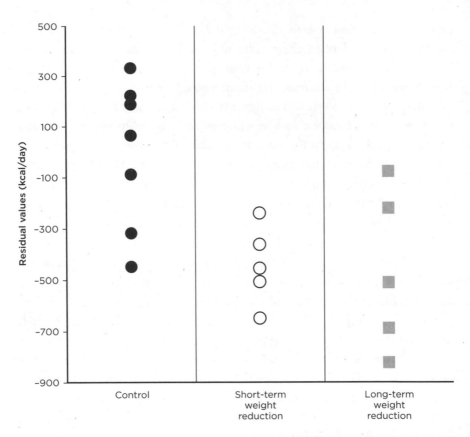

Figure 4.2. Subjects in a study of the effect of a reduced-calorie diet showed slower metabolism (in terms of total energy expenditure) compared to the control group, in both the short term and the long term. Adapted from Michael Rosenbaum et al., "Long-Term Persistence of Adaptive Thermogenesis in Subjects Who Have Maintained a Reduced Body Weight," *American Journal of Clinical Nutrition* 88, no. 4 (October 2008): 906–12.

we need to cover our energy usage until the next meal. The subjects in the Minnesota Starvation Experiment spent a lot of time in the less-fed state because they were eating each day, but consuming less food than needed for health and vitality. Spending time in the less-fed state seemed to strengthen their body's determination to prepare against food scarcity in the future.

Another study leads to a similar conclusion. Researchers at Rockefeller University and Columbia Presbyterian Medical Center studied subjects who had lost weight by being on a controlled reduced-calorie liquid diet for months at a time in an in-patient facility. We might assume they consumed the diet on a typical meal schedule since no special mention was made otherwise. Thus, they were put into the less-fed state for a prolonged period of time. The subjects lost weight, (see figure 4.2) and their metabolisms slowed down and *remained slow* for over a year after the reduced calorie diet had ended.[12]

The researchers compared the total energy expenditures of the subjects, dividing them into three groups: (1) people who hadn't participated in the weight-loss program (the control group), (2) people who participated in the program and who maintained a weight reduction of greater than or equal to 10 percent for five to eight weeks (short-term group), and (3) people who participated in the program and maintained a weight reduction of greater than or equal to 10 percent for over a year (long-term group). They found that the reduction in metabolic rate in the third group (after a full year) was similar to the reduction of metabolic rate in the second group (after a couple of months). This suggests that the subjects' bodies had been trained to store fat in preparation for possible food scarcity. Their bodies begrudgingly gave up their energy stores when they were less fed and they were left with a determination to regain those energy stores even a year later. In other words, their set points increased.

These studies show some of the effects of spending more time in the less-fed state. However, spending more time in the non-fed state has very different effects. To help see how different they are, let's go back to the budget analogy. Suppose you are quite frugal and diligent about building up your savings for a rainy day. You scrimp to put every available dollar into your savings. After many years, with careful saving and fortunate investing you amass ample savings, which gives you a great feeling of security. Then a rainy day comes, and your income is cut in half. You have become so accustomed to storing money that you decide not to touch your savings but instead to cut your expenses. You buy less expensive food, clothing, and shelter, but that doesn't serve you well, and your health and earning ability soon begin to fade. You are experiencing the very kind of rainy-day deprivation that you had prepared for so carefully. Did I mention that your rainy-day savings had grown into the millions? So you reevaluate and decide to take a leave of absence from work. You start getting plenty of rest, eating healthy food, and exercising. You even travel to see the world as you've always wanted. Wait a minute. When your income was cut in half you struggled desperately, but when your income dropped to *zero* you sprang back to life? Yes, because you decided to live off your *storage*, not your *income*. Once you regain your health and vigor, you can always go back to work if you want, perhaps with greater earning potential, to again prepare for your future.

The savings account in this analogy represents your stored fat, and burning off your excess fat can be very good for your health. One way to signal the body to burn stored fat is to spend more time in the non-fed state—with *zero* income of food. That doesn't mean never eating, it means alternating between the fed state and the non-fed state (which starts when you feel hunger) and pushing off your hunger with fastercise instead of allowing yourself to go around hungry and miserable in the less-fed state, as many people do when they're dieting. This simple yet profound difference can help you remain comfortable while you give your body a break and spring back to life and regain your health.

At the beginning of the recovery phase in the Minnesota Starvation Experiment, the lead researcher gave the study subjects some vitamin and protein supplements without increasing their calories very much to see if it would make much difference. He found that it made no difference at all. The only thing that refreshed them was giving them more food (energy), which he did three weeks after starting the vitamin and protein supplements. Energy fuels life. Fortunately, we have plenty of energy in our fat stores; we just need to know how to access it.

"My doctor encouraged me to use fastercise at times that I was most vulnerable to snacking and eating food items not on my health management plan. I already had lost forty-five pounds and still wanted to maintain my weight loss and lose another fifteen pounds. I am happy to say that I am down another ten pounds with the use of fastercise. It has helped with cravings during my busy days."

—CARLENE S.

The non-fed state is hormonally conducive to fat loss. We get to decide how much time we spend in the fed state (which includes the less-fed, normal-fed, and more-fed states) and how much time we spend in the non-fed state. Do we want to spend more time in the fed state or more time in the non-fed state? What's normal? The definition of *normal* is "conforming to a standard that is ordinary, typical, customary, conventional, habitual, accustomed, or expected." The behavior that is conventional and habitual presently in the United States is generating epidemic levels of overweight and obesity. As mentioned earlier in this chapter, it's likely that in the past, when obesity was less rampant, it was normal to spend more time in the non-fed state than we do today.

I believe many people would benefit from dialing back a little on their time spent in the fed state. Going a little longer without food than we currently do while putting off hunger with fastercise can still fit easily into our lifestyle of eating with friends and family. There are many different ways to schedule more non-fed time into our lives. The amount of time you spend in the fed state depends on the size and frequency of your meals. The bigger the meals and the more often you eat, the more time you spend in the fed state because it takes the body longer to process a bigger meal.

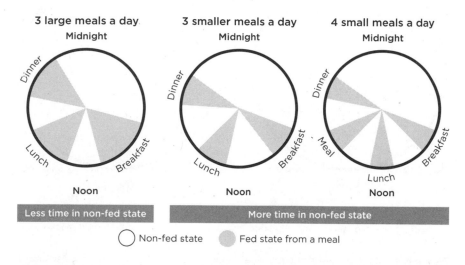

Figure 4.3. If we eat smaller meals with the same frequency, we will spend more time in the non-fed state. And if we eat less total food in a day but it's spread out over a greater number of more frequent meals, we might still spend more time in the non-fed state.

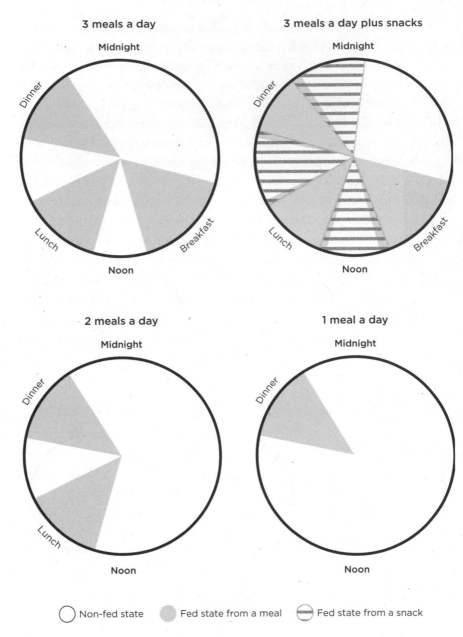

Figure 4.4. When we eat three meals a day without snacks, we spend most of our time in the non-fed state. On the other hand, when we eat three meals a day and grab snacks in between, we can easily spend almost all of our waking hours in the fed state. Changing to a two-meals-a-day schedule or a one-meal-a-day schedule results in significantly more time spent in the non-fed state.

Different combinations of meal size and frequency add up to different amounts of time in the non-fed state. If you eat smaller meals with the same frequency, you will spend more time in the non-fed state. And, if you eat less total food in a day but it's spread out over a greater number of more-frequent meals, you might still spend more time in the non-fed state, as shown in figure 4.3.

You can also schedule in more time in the non-fed state on a daily, weekly, or monthly basis. And you don't have to stick with the same schedule forever, either. Once you reduce your fat stores, you can always readjust your eating and exercise habits. Figure 4.4 shows how changing your daily meal schedule can allow you to spend more time in the non-fed state. And in chapter 6, I describe specific ways to use fastercise to help you make this kind of schedule change without having to endure the discomfort of feeling hungry when you're in the non-fed state.

If you ask some of your friends or acquaintances who are lean and fit about when and how much they usually eat during the day, I think you'll discover, as I have, that they tend to spend more time in the non-fed state than what might be considered normal. But to them, the way they eat is normal! They may say something like, "I'm a big breakfast guy, and I often don't eat lunch, and I often eat very little dinner." Or, "I don't usually eat breakfast or much lunch, but I do eat dinner." They don't think of themselves as being on a special diet. That's just how food fits in for them. They don't seem very preoccupied with food.

Some people choose to go even longer periods without food, getting into ketosis after two or more days without eating. Each week, they may choose one thirty-six-hour period to go without food (taking only water and noncaloric drinks), and once every three months, they may do a three-day fast. Once into ketosis, some people experience improved cognitive function and better productivity, which is not surprising, because for centuries, fasting has been practiced for cleansing and rejuvenation and improved health and well-being.

Benefits of the Non-Fed State

By now, I think you're getting the message that the non-fed state is as important for our health and well-being as the fed state is. It's also important to maintain a healthy balance between the two. This means that if we are spending too much time in the fed state, perhaps one of the best things we could do for our health is to spend more time not eating!

As I researched the benefits of the non-fed state, I was surprised to learn all the ways in which it can contribute to better health. Here's a summary of some of the most important benefits.

Improved Digestive Health

One of the core principles of natural healing is "all disease begins in the gut." More and more research is revealing just how important proper gut function is. Like our skin, the gut separates us from the outside world. But unlike the skin, the gut is more like a filter that is built to allow some of the outside world (for example, food particles) to enter the bloodstream, while keeping out the rest. Modern research is showing how poor gut function can contribute to autoimmune disease and other health problems. Spending time in the non-fed state gives our gut a rest and a chance to recuperate.

Relief from Allergy Symptoms

Many people who suffer from allergies notice a dramatic reduction in their allergy symptoms when they began eating less often. Sometimes their symptoms resolve completely within five to seven days of reduced eating—that is, spending more time in the non-fed state. How might this be? Fundamentally, an allergy is a condition in which the immune system reacts abnormally to a foreign substance. The gut harbors the largest number of immune cells in the body.[13] The gut is tasked with taking in food while keeping out intruders and toxins. When the gut is in disrepair and becomes leaky, many problems with immunity can occur (including autoimmune conditions such as Hashimoto's thyroiditis). In many cases, gut function and therefore immune function can improve significantly just by giving the body more of an opportunity to rest and recover and heal. It may be that many of us are simply spending too much time in the fed state, which is causing more wear and tear in the gut and less chance for repair.

Weight Loss

A recent study at the University of Illinois at Chicago compared spending more time in the less-fed state with more time in the non-fed state and the impact of each scenario on weight loss. The researchers found that subjects who lost weight while spending more time in the non-fed state lost less muscle mass.[14] Another study showed that obese patients adapted well to spending more time in the non-fed state and were able to lose significant amounts of weight without feeling very hungry.[15] Another study showed that epinephrine (which is increased by fastercise) is much more effective at burning fat when people are in the non-fed state.[16]

In chapter 3 I explained the importance of growth hormone (GH) in mobilizing fat stores while preserving muscle and how, for our purposes, this is a very valuable combination. The levels of GH are frequently *undetectable* in the fed state. GH is secreted in pulses. More time in the non-fed state significantly increases the number of pulses, the amplitude of the pulses, and the total amount of GH secreted. Fastercise also increases GH. Thus, we can see the additive benefit of answering hunger with fastercise.

Relief of Menopausal Symptoms

Spending more time in the non-fed state can help dramatically with problems people wouldn't normally associate with the choice of how often to eat. For example, many women notice that their menopausal hot flashes are much worse after they eat and are greatly reduced or eliminated when they spend more time not eating. Many women also tend to gain a little weight during menopause, even if they haven't changed their total caloric intake, apparently due to a slowing of the metabolism, which fastercise may help reverse. It may be that as weight increases, the body notices that the weight set point has been exceeded, and it may try to reduce weight by increasing energy expenditure in the form of a hot flash. This could explain why more time in the non-fed state might reduce hot flashes.

Better Brain Function

Research has also demonstrated that more time in the non-fed state greatly encourages our brain cells to refurbish themselves through the process of autophagy,[17] which is the natural, regulated destruction and recycling of unnecessary or dysfunctional cells and cellular components. Critical for good health and prevention of chronic disease, autophagy is known to be triggered by intermittent fasting and is thought to be one reason fasting appears to stave off aging. Yoshinori Ohsumi was awarded the 2016 Nobel Prize in Medicine for his discoveries of mechanisms for autophagy. Spending more time in the non-fed state may help our brains age better and help us live longer.[18] The non-fed state increases brain-derived neurotrophic factor (BDNF), which is a protein that increases the resistance of brain cells to dysfunction and degeneration.[19] Ketones also increase in the non-fed state and can have many beneficial effects, which I discuss more in chapter 9.

Improved Immune System Function

High blood sugar levels can contribute to oxidative stress and inflammation that leads to cardiovascular disease.[20] Spending more time in the non-fed state reduces such inflammation by decreasing pro-inflammatory cytokines and immune cells.[21] Another term for abnormal inflammation is *immune dysfunction*. Autophagy and immune dysfunction come together in a study done by researchers from the University of Southern California that showed that just three days of fasting can cause significant decreases in old white blood cells and significant replenishment with new white blood cells (important in immunity) upon refeeding.[22] These results demonstrate how more time in the non-fed state can lead to a significant reboot of our immune system, which may help us overcome immune-related problems (including damaged immunity secondary to chemotherapy).

Heart Health

Inflammation contributes to cardiovascular disease such as increased blood vessel wall thickness and atherosclerosis. Caloric-restriction studies have shown reduction in blood vessel wall thickness of the carotid arteries as well as improvement in cholesterol and triglycerides.[23] Multiple studies show that time in the non-fed state can reduce inflammation and high blood pressure.[24] In one study, 174 people with hypertension were treated with a water-only fast for about eleven days. Almost 90 percent of the subjects achieved normal blood pressure. The average blood pressure drop in patients with severe hypertension was 60 points systolic. All of the subjects who were taking blood pressure medicine at the start of the study (6.3 percent of the total) successfully discontinued the use of the medicine.[25] Just as BDNF is helpful in preserving brain function, it appears to also be helpful in helping people survive congestive heart failure (CHF). More time in the non-fed state increases BDNF and increases survival of CHF.[26]

Longevity

The importance of a healthy fed/non-fed balance brings with it the notion that not eating can be key to thriving for the long term. Longevity has also been attributed to eating less. Apparently, we are not built to have free access to food all the time. When laboratory animals are given free access to food, they tend to gain weight and die younger.[27] Horses are grass-eaters by nature, but when they break into storage bins of sweet feed (grain), they have been known to eat themselves to death. In a sense, many of us are also eating ourselves to death. Luigi Cornaro (1467–1566), a Venetian noble-man who suffered from poor health, abandoned a life of overindulgence and eventually wrote a book about how he and others could live long and active lives by eating less. Cornaro lived to almost one hundred years of age, which is a long life even by today's standards. Interestingly, animals that are put on a calorie-restricted diet can live 15–20 percent longer than similar littermates without having abnormally low body mass.[28]

Intermittent fasting appears to increase lifespan via its effect on mito-chondria.[29] Research shows that the enzyme AMP-activated protein kinase is known for protecting mitochondria.[30] It is one signal that exerts control on the aging process. It balances the metabolism, enhances stress resistance, promotes good cellular housekeeping, and is stimulated by calorie restric-tion and impaired by overfeeding.[31] Active autophagy is critical in fighting age-related disorders and improving healthspan, and drug companies are looking for drugs that can activate the autophagy pathways.[32] It's wonder-ful that those pathways can be stimulated simply by spending more time in the non-fed state. Not only may calorie restriction have a life-prolonging effect but it may also be promising in the prevention of cancer.[33]

CHAPTER 5

Finding the Right Balance

I t can be very rewarding to set a goal, make a plan, work your plan, and accomplish your goal—especially when you've selected a rewarding goal. But imagine spending a great deal of time, money, planning, and effort to climb a mountain, only to realize that you've climbed the wrong mountain. Sometimes people who set a goal of losing weight spend a lot of time, money, and effort eating less and exercising more. And some then lose weight, but others gain weight. Some people keep the weight off, but others end up regaining it. I suggest that you'll see more consistent results if instead of setting a goal of "lowering your weight," you set your sights on "lowering your weight set point." In this chapter, I present a new paradigm that explains how we can climb this mountain by aligning ourselves with the body's survival instincts.

As we learned in chapter 4, the position of your weight set point determines your tendency to lose weight and keep it off. What we weigh is made up of fat, muscle, bone, organs, blood, and more. However, when people want to "lose weight" they usually mean they want to lose fat. Fat is our largest energy store and I believe the weight set point is a reflection of how much energy our bodies are inclined to store in the form of fat. Therefore, I use the terms *fat set point* and *weight set point* interchangeably. We tend to focus on losing fat, but what good does it do to lose fat in a way that increases the fat set point? If we had a hard time keeping fat off with our previous set point, it will be even harder when our set point has increased. Clearly, we should focus on lowering the body's fat set point. Focusing on lowering the fat set point will help us lose fat, and that fat will tend to stay off. That's the way to go.

A study published in the *New England Journal of Medicine* in 1995 also supports the concept of a weight set point. Researchers measured the

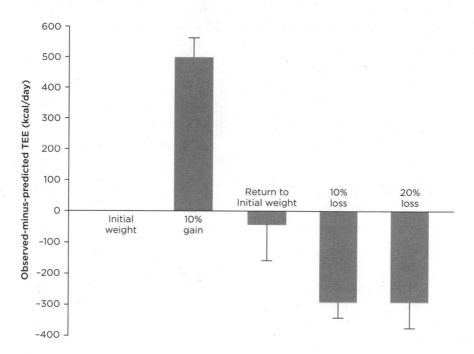

Figure 5.1. Metabolic rate, measured as observed minus predicted total energy expenditure (TEE), increased in research study subjects who gained weight and decreased in those who lost weight. Adapted from Rudolph Leibel, Michael Rosenbaum, and Jules Hirsch, "Changes in Energy Expenditure Resulting from Altered Body Weight," *New England Journal of Medicine* 332, no. 10 (March 1995): 621–28.

changes in total energy expenditure both in patients who gained body weight and in those who lost body weight. Total energy expenditure of each participant was measured at both the beginning and the end of the study. One group of participants had gained 10 percent of their body weight, one group had lost 10 percent, and a third group had lost 20 percent.

The researchers found that total energy expenditure rose for the group that gained weight and decreased in the groups that lost weight, as shown in figure 5.1.[1] Thus, when the subjects went above their normal weight, their metabolism sped up, seemingly in an effort by the body to get back to original weight. By the same token, the decrease in metabolic rate in those who went below their normal weights also seems to be an effort to return to original weight.

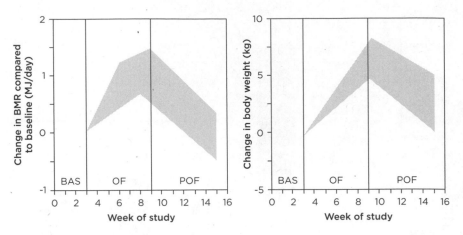

Figure 5.2. These study results show that metabolic rate and weight increased with overfeeding and then declined when overfeeding was discontinued. *BMR* stands for basal metabolic rate, *BAS* stands for baseline, *OF* stands for overfeeding, and *POF* stands for post overfeeding. Adapted from Erik Diaz et al., "Metabolic Response to Experimental Overfeeding in Lean and Overweight Healthy Volunteers," *American Journal of Clinical Nutrition* 56, no. 4 (October 1992): 541–55.

Another study involving six lean and three overweight young men demonstrated that basal metabolic rate and weight went up with overfeeding and went back down when overfeeding was discontinued, as shown in figure 5.2.[2]

Working with the Body's Priorities

In chapter 1, I presented the body's four basic priorities (avoid passing out, being killed, damaging your muscles, and starving). Avoiding starvation has been a prime directive for all life from inception. Failure at this means death. Humans have developed two approaches to meet this priority. Historically, humans foraged for food on a regular basis, hunting and gathering. Since there was no refrigeration and little means for storage, foraging for food was a daily part of life. Thus, our earliest ancestors were foragers. The other approach developed only very recently in human history: agriculture. Roughly twelve thousand years ago societies began to domesticate plants and animals. In a way, this was the beginning of the storage of food. Food wasn't stored in refrigerators or pantries but rather in flocks and fields. In the last one hundred years, our capacity for producing and storing food has grown like never before. Over 99 percent of American homes have at least one refrigerator and the average size of refrigerators has continued to grow over time. The food of most consumers isn't stored in their own flocks and fields but in grocery stores, fast-food restaurants, refrigerators, and kitchen cabinets. Modern conveniences have snowballed since 1900, and so has obesity. For eons humans have been built to be foragers, almost always needing to exert ourselves in some way whenever we were hungry and before we could eat. Today, due to our food production and storage technology and access, we can just walk into the kitchen or pull over to a drive-through window and procure a full meal with little to no exertion.

The two strategies for avoiding starvation—*foraging for food* and *storing food*—apply not only to us as a society but also in our individual bodies. To avoid starvation, the human body can exert muscular effort in order to forage new food, or it can turn to the energy stored in kitchens and restaurants and in the body's storage depots of glycogen and fat. Therefore, I refer to one survival strategy as *forage mode* and the other as *storage mode*.

Another of the body's four basic priorities is to avoid being killed by a predator, an enemy, or a dangerous situation. To escape danger humans need speed, reflexes, muscular strength, and endurance. These are the same resources needed for foraging. Therefore, I include escaping danger in the forage mode.

The Survival Balance

Just as both income and savings are helpful for financial well-being, being able to forage for food and also to store fat are helpful for our survival. However, storing more fat and building less muscle makes it harder to move quickly and easily while foraging. And building more muscle to forage and storing less fat makes it harder to survive a famine. Thus, it's easy to see that surviving and thriving depend on a healthy balance between the two strategies of forage and storage. This balance is what I call the *survival balance*.

The survival balance helps explain why our hunger goes away after a meal, even though more food is available and we can still fit more in our

Figure 5.3. The less muscle and the more fat our bodies have, the less likely we are to starve in circumstances of prolonged lack of food. But the more muscle and the less fat our bodies have, the more easily we can obtain food and escape from dangerous situations. The survival balance is the balance between these two strategies—storage and forage.

stomachs. It also explains why we tend to eat less and lose weight after a period of overfeeding even when we have free access to food: so we can forage more easily. I believe that your body's fat set point and your muscle mass are largely determined by your body's survival balance between the forage mode and the storage mode. When you shift your survival balance, you are shifting your fat set point. The more the body is focused on storage, the more fat it tries to store. The more reliably successful you are at foraging a meal, the less fat your body needs and wants to store so you can continue to forage successfully. The more vigorously you fastercise, the more you signal the body that the foraging is challenging and that you need muscle in order to succeed. And when you always fastercise before you eat, your body receives the signal that your foraging skills always keep it from going hungry and there's no need to store excess fat.

The survival balance is manifested plainly in how your body is built and how it functions. On a cellular level, the cytosolic process also helps us avoid starvation by delivering energy as fast as possible to help us forage more successfully. The mitochondrial process (described in chapter 2) supports the body's priority to avoid starvation by extracting as much energy from our food as possible. This helps put the most energy into and get the most energy out of our *storage*. On a hormonal level, the counterregulatory hormones support the forage strategy by helping us liberate as much energy as quickly as possible from our energy stores. Insulin supports the storage strategy by helping us store as much energy from our food as possible.

Historically, people under conditions of food scarcity surely benefited from shifting their survival balances toward storage. This happened automatically because they ate less and expended less energy foraging—there simply wasn't food available. On the other hand, when wild game and fruits and berries were plentiful or if they spent more time fending off predators or enemies, then they benefited from their survival balances shifting toward forage mode. This happened automatically when they engaged in vigorous muscular action (to obtain food) before eating.

Shifting the Balance

Shifting the survival balance in either direction takes work. The body resists going toward storage mode because it knows it also needs to be able to forage and flee. And it resists going into forage mode because it knows that being able to store energy is also important.

It appears that the non-fed state can help move the survival balance more toward the forage mode and toward the lowering of our fat set point, especially when coupled with fastercise. This might seem counterintuitive. After all, the studies I discussed in chapter 4 showed that reducing food intake slowed the metabolism and shifted the balance toward storage mode. On the other hand, the metabolic rate actually increases after thirty-six hours without food.[3] Why would this happen? Why does the metabolism slow down when we eat *fewer* calories more often and speed up when we go longer with *zero* calories?[4]

The distinction is that the research study subjects were spending a lot of time in the less-fed state, with its higher insulin levels, not more time in the non-fed state. Therefore, their bodies remained focused on food as their fuel source.

Reducing the quantity of that fuel shifted the survival balance toward storage mode, leading their bodies to store as much energy as possible. In contrast, if we choose to spend more time in the non-fed state (with very low insulin levels) and fastercise, we are directing the body to rely more on stored fat for fuel and less on food we have on hand. It's the difference between *reducing* our fuel source and *switching* our fuel source.[5] That's a huge difference. When the body turns to burning stored fat, it sees the result as an increase in available energy, because the amount of energy available in stored fat is immense, which can lead the metabolism to speed up.[6]

We've all had the experience of being told how *not* to do something without being told how *to* do it. We can be left with a feeling of uncertainty and indecision. Or there may be times when we know what we don't want to do but haven't yet figured out what we do want to do because we have so many options (like high school seniors trying to decide what to do with the rest of their lives). It can take time to decide. When left to itself, the body tends to keep its options open so that it can respond to an almost endless variety of circumstances. Thus, when we enter the non-fed state without answering hunger with fastercise or eating, it is like telling the body there's no food on hand to eat without telling it what to do about that. The body wonders whether it should start burning stored energy, or whether to conserve because food might be scarce. The longer we remain in this state without eating, the more the body transitions to stored fuel. The key word here is *transition*. It's a gradual transition marked by hours of hunger and discomfort. Hunger is meant to get your attention. The body really wants to know how you want to handle this urgent situation. Do you want to focus on conserving energy and building fat stores, or do you want to focus on foraging for food? If you want to promote fat storage, then don't fastercise before eating; if you want to promote muscle growth, then always fastercise before you eat. When you eat without tightening your muscles soon beforehand, you are signaling your body that you are relying on the storage strategy, that there's stored food on hand, and that you're focused on building your body's fat stores. When you answer hunger with fastercise, it's like telling your body to burn stored fuel to cancel your hunger so you can more easily forage for more food.

Fastercise Shifts the Balance

Fastercise simulates the energy usage of activities related to immediate survival, such as escaping a dangerous situation, hunting down prey for food, gathering food from hard-to-reach places, or shivering to keep warm. When we vigorously tighten our muscles every time we get hungry and before each meal we eat, we send our body the signal that we are taking action to obtain food. Our muscular action stimulates the release of stored fuel and the body is satisfied for a time and hunger goes away for a time. We may push off hunger a number of times, and as long as we're taking action the body is satisfied. However, at some point we can decide it's time to eat again, especially when the hunger signal becomes insistent. Nevertheless, before we eat we can fastercise one more time to signal the

"Fastercise seemed unnatural at first because it was a new activity. Exercise can sometimes stimulate my appetite, so I was skeptical. I believe it helped, though, as the short bursts of activity it provided did help with hunger. I am still playing with the length of activity to see what will stimulate my hunger and what will help ease my cravings."   —Corrine L.

body that we are foraging and not eating stored food already on hand. Keep in mind that throughout here, when I talk about muscular action and tightening muscles, I'm not talking about a sweaty thirty-minute workout; I'm talking about fastercising for a matter of seconds.

Of course, in our modern society food is usually readily available all the time without our having to physically exert ourselves at all. Because of this great convenience there is nothing to drive us to contract our muscles when we get hungry and before we eat. We need to do it intentionally if we want to shift our survival balance more toward leanness.

It may sound strange, but some forms of continuous exercise, like riding a bike or running on a treadmill, may not send a strong forage signal, especially if not done soon before eating. Although this type of exercise uses up calories and can help us perform well in athletic competitions, it doesn't strongly simulate foraging, because it doesn't involve a fast rush of intense activity. Instead, this type of exercise can signal the body that you're looking for food, but food is scarce. This can focus the body on conserving energy and increasing your appetite (we call this "working up an appetite") to remind you to take in plenty of food to fuel your migration to a place where foraging is more successful. Thus, cutting back on calories and "migrating" for long distances on a treadmill without eating soon after can promote your sensation of hunger and may even shift the survival balance toward storage mode, increasing the fat set point. Thus, this survival balance paradigm explains how we may be able to lower our set point more with one minute of high-intensity exercise that simulates foraging than we can with an hour of lower-intensity exercise. I think a good rule of thumb is that if you are having trouble getting rid of fat, you are probably fighting against, rather than working with, your body's priorities and hormones.

We each depend on a combination of acquiring and storing food in order to live. There is a balance between the two strategies of hunting-gathering and storage, and we can signal the body how much to rely on each. Where is the balance between the signals that we send our bodies in our modern society of convenience? When you exercise three times a week, you are exercising before three meals. However, many people eat twenty-one meals a week. By tightening your muscles at least before every meal you could send the hunter-gatherer signal to your body seven times more in a week. That's a huge difference.

Feeding States and Survival Balance

As explained in chapter 4, the more-fed state and the less-fed state are both components of the fed state. Significantly, it appears that spending more time in either the more-fed or the less-fed state tends to shift the survival balance toward storage mode, increasing set point. Cutting way back on calories, and eating small, even healthy snacks every few hours to manage your hunger may leave you for some time in the less-fed state. This could increase your fat set point because your insulin level increases whenever you eat, and insulin drives the body toward storage. So, the common question dieters ask is "Why haven't I lost fat when I'm eating so little?" But if we rephrase the question, it gives us a different insight: "Why haven't I lost fat even though I'm eating all the time?"

In another study that demonstrates this point, the appetite hormones that encourage weight gain increased after weight loss on a less-fed program and were elevated even after a year.[7] In this study, researchers enrolled fifty overweight or obese patients in a ten-week less-fed weight-loss program using a very-low-energy diet. Patients were fed a calorie-restricted diet on a normal schedule of three daily meals. Researchers measured levels of the hormones that affect appetite before the diet was implemented, at the end of the ten-week diet, and a year later at sixty-two weeks. They measured the levels of leptin, ghrelin, and others along with subjective ratings of appetite. They found that the hormones that increase appetite, and appetite itself, were increased by the end of the diet and *remained elevated* a year later. Putting the study subjects on a less-fed diet shifted their survival balances even more toward storage mode, directing their bodies to acquire and store as many calories as possible to avoid starvation.

I explained earlier how overfeeding has consistently been shown in animals to increase the weight set point; this probably occurs in humans as well. It appears that the more time we spend in the fed state, the more we signal our bodies to depend on food for energy and to increase our storage and set point. The fed state aligns with storage mode because it facilitates storing energy from food. Thus, we can shift the survival balance toward storage mode and increase our set point by:

- Spending more time in the fed state
- Spending less time foraging

Or we can shift the survival balance toward forage mode and lower our set point by:

- Spending more time in the non-fed state
- Spending more time foraging

The non-fed state aligns with the forage mode because it signals the body to tap into the body's energy stores.

Responding to the Hunger Signal

As explained in chapter 3, insulin is the switch that determines whether the body engages in utilizing energy from food or from storage (glycogen, muscle, fat).[8] Low insulin levels are crucial for easy access to fat stores. While most people trying to lose weight dread the onset of hunger, it's actually an exciting moment. The onset of hunger is the moment our fat stores become available for burning and is the beginning of losing fat, which is what we want! Of course, we can quickly put that stored-fat burning to an end by eating something to make our hunger go away. Or we can encourage that stored-fat burning by using fastercise to direct our bodies to snack on storage instead. When we cancel our hunger with fastercise, we are doing it by feeding our bodies with the energy from stored fat and glycogen. The more time we spend burning fat, the more fat we lose. That's incentive to eat smaller meals, so we can get hungrier sooner so we can resume snacking on our storage. Leaner and leaner every hour without going hungry? What's not to like? Fastercise allows us to avoid discomfort (feeling hungry) in the non-fed state. Then we can enjoy our meals in the fed state, knowing we will not have to bear with hunger pangs the next time we enter the non-fed state. We have the fun of losing fat *and* the fun of enjoying food, not to mention the strength and vitality we gain from a healthy survival balance.

Even if you don't fastercise, remaining in the non-fed state signals your body to gradually transition its fuel supply from food to storage over a couple of days. At first, it is waiting to determine as hours pass whether it is in the less-fed state (which raises set point) or the non-fed state (which lowers it). I refer to this period of energy deprivation as the *hunger zone*, which is when hunger and muscle loss is most intense. As complete caloric deprivation continues, your body determines that you are ignoring all the warning signals of hunger it's sending. At that point, your body enters

The Hunger Crossroads

When you become hungry, you have three options.

1. If you don't eat or fastercise:
   - Stored fuels are burned for fuel, including muscle.
   - Metabolism slows at first.
   - Hunger can come and go and vary in intensity.
2. If you eat:
   - Food is burned for fuel.
   - Metabolism doesn't slow.
   - Hunger goes away in minutes, and returns later.
3. If you fastercise:
   - Stored fuels are burned for fuel and muscle is preserved.
   - Metabolism doesn't slow.
   - Hunger goes away in seconds, and returns later.

ketosis and begins relying mostly on stored fat for fuel. Your hunger and muscle loss decrease significantly, though some muscle loss still occurs. This complete dependence on storage instead of food can shift your survival balance away from a dependence on food and lower your set point. Not only does traversing the hunger zone without fastercise take time, but it also requires dedication and discipline because it can be unpleasant. Fastercising periodically when you are in the hunger zone can eliminate a lot of the discomfort and difficulties associated with fasting.

Unintentionally fighting against the body's priority to store and conserve as much energy as possible can lead to confounding results. You eat less and the body conserves; you exercise more and the body conserves. It's as though there's no way out. What on earth can you do to persuade the body to work with you?

You tell your body, "I want to lose fat."

Your body responds, "Give me one good reason I should spend energy rather than store it."

You can respond, "To obtain food and to escape a dangerous situation."

And the body replies, "Sure thing, that's what the energy is there for."

Wow, what a difference! The need to obtain food and escape are perhaps the only reasons the body will freely spend energy. After all, it's all about survival. It can make a big difference when you focus your eating and exercise strategies to shift your survival balance away from storage mode and toward forage mode.

Spending time in the less-fed state promotes the building of fat stores and the loss of fat stores *at the same time*. This sounds impossible, but it's not, and understanding this key concept can help alleviate a huge amount of confusion and frustration people experience when trying to manage their weight. Being in the less-fed state, especially if you're not fastercising before you eat, tends to move your survival balance toward storage mode and increase your set point because you are training your body to rely on only the scarce food on hand. It increases the body's *tendency* to gain fat, or in other words, its desire to gain fat. At the same time, spending time in the less-fed state can result in some fat loss in the short run. When we lose fat but simultaneously increase our body's desire to conserve and build fat stores, we tend to gain all the fat back over time and then some, as did the subjects in the Minnesota Starvation Experiment.

Shifting Your Life Balance

The survival balance concept is essentially the balance between the two approaches we have for avoiding starvation: storage and acquiring new food. Or, in other words, it is the balance between storage and income. This balance applies not only to societies, and our own bodies, but also to how we manage our lives. It applies to societies in many ways. We've already discussed how our hunter-gatherer ancestors relied on acquiring food to support themselves and how, more recently, agrarian societies have emerged relying on food storage. Our bodies can shift toward storage mode or forage (income) mode. Storage is good and income is good. It's great to have a healthy balance, and we each get to decide what balance works best for us in our present circumstances.

For example, from a financial standpoint, it's good for us to store (save) money that we can use in the future, and it's also good for us to increase our income if we can. Sometimes, we need to spend some of the money we've saved in order to get training or make an investment that may help us to increase our income. Similarly, it is good to keep some of the possessions that we've acquired in the past because they may be useful to us in the future. However, sometimes we keep baggage (both literally and emotionally) that is unlikely to serve us well in the future. Clutter results from relying heavily on storage of the past rather than on income for the future. However, when we focus more on the future, we can let go of storing things we no longer need and we can organize what's left so that we can more easily utilize it to thrive in the future. Clutter comes from the past; organization is about the future. The balance between storage and income applies universally and it applies in our own bodies as well. We get to decide if we want our survival balance to shift more toward storage or more toward forage.

When deciding how to manage your weight, you could focus solely on what you eat and how you balance your time between the fed and non-fed states. Or you could focus solely on the signals you are sending with your physical activities whether to store or forage. Either could lead to some unexpected and disappointing results, however, because you might end up sending your body mixed messages. For example, fastercising while in the fed state may not direct your body to burn much stored fat, because the meal you just ate has increased your insulin level, signaling your body to store fat. However, by canceling hunger with fastercise you can send your body a powerful combined signal to shift your survival balance toward forage mode and leanness.

Answering hunger with fastercise puts us back in sync with the built-in priorities that our bodies have inherited from our foraging forebears. Right now, many of us have a sedentary lifestyle and follow this pattern: eat, get hungry, eat, get hungry, eat, get hungry. This pattern favors only storage mode. Here's the pattern that favors a healthy survival balance: eat, get hungry, forage, eat, get hungry, forage, eat, get hungry, forage. This pattern sends our bodies the message that we must preserve and build muscle, because we need our muscles in order to obtain food. This pattern gives our bodies a reason to move the survival balance more toward forage and away from storage, more toward muscle and away from fat.

CHAPTER 6

Getting the Most from Fastercise

P icture a marathon race where two prizes are given: one for the runner who crosses the finish line first, and one for the runner who achieves the highest speed during the race. Obviously, these are two different things, and each one is noteworthy in a different way. Similarly, there are unique benefits from reaching top speed with one minute of fastercise that you can't achieve by doing long stints of lower-intensity exercise. And many of us would rarely, if ever, reach the peak exercise intensity of fastercise while doing some other form of exercise instead.

As we've discussed, fastercise signals the body to mobilize glycogen and fat stores, which releases a tremendous amount of fuel and energy that cancels out hunger pangs and preserves and possibly builds muscle. When our bodies mobilize glycogen and fat and make ATP, about 66 percent of the energy stored in the glycogen and fat is released as heat. Then, when our bodies use that ATP to generate muscle power, about 66 percent of the energy stored in the ATP is also lost as heat. That's a lot of heat released! And generating heat requires a lot of energy. The appliances in our homes that use the most electricity are the ones involved in heating and cooling, such as the water heater, furnace, refrigerator, and air conditioner. In my experience, one sixty-second bout of fastercise is enough to raise my body temperature one degree Fahrenheit, or more. By my calculations, that's enough heat to boil about three cups of water! Now, I'm not saying that pedaling a stationary bike powering a heater would boil water in one minute, because most of the energy expended would go toward heating our own bodies. And I'm not saying that exercise alone is enough to generate a lot of fat loss.[1] But I can tell you that fastercise is fantastic for helping burn fat, preserve muscle, and remain comfortable as you spend more time in the non-fed state. Within one minute, high-energy exercise such as fastercise can increase energy consumption 120-fold from baseline, and this immediately stimulates the breakdown of stored fat.[2]

This chapter is full of practical information about how to implement fastercise in your daily life to get the maximum benefit. Your diet sends signals to your body about whether to store fat or burn it, slow your metabolism or speed it up, build muscle or break it down. Your physical activity sends signals about all those things as well. It's important that you coordinate the signals you are sending your body with your diet and exercise because sending mixed messages can generate puzzling and frustrating results. In this chapter, you'll learn how to time your diet and exercise according to your appetite in order to work with your body to accomplish your goals in an extremely time-efficient and comfortable manner. You'll learn tips and variations of tightercise and shivercise and when to do each, as well as health benefits of fastercise other than fat loss and muscle building, and how even people with physical limitations can do fastercise.

Swinging with Meals

The old saying "timing is everything" reflects how doing the right things at the right times can make a huge difference. An example I like to use is swinging on a swing. When you first sit down on a swing and try to get it started from a dead stop, it can be difficult to gain momentum. I'm not talking about having someone push you, or even using your feet to push off against the ground. No, I mean you sit on the swing with your feet dangling and you try to get yourself swinging. In that scenario, you must time your actions just right to initiate the swing's movement. If you lean the wrong way at the wrong time, you end up canceling out your progress. Once you gain some momentum, it's much easier to coordinate your actions properly, because you have a bigger window of opportunity within which you can successfully take each action. In a similar way, you can time fastercising between meals to build momentum in the hormonal swings that contribute to the burning of fat when you're in the non-fed state and the building of muscle when you're in the fed state.

When we eat a regular-size meal and answer hunger with one or more bouts of fastercise, as in figure 6.1, we are using fastercise to propel ourselves through a significant chunk of time in the non-fed state by mobilizing glycogen and fat while preserving our muscle and metabolic rate. On top of that, by spending more time in the non-fed state, we increase our insulin sensitivity, which means that our muscles will benefit more the next time we do eat. Fastercise helps us lower our glycogen stores so that when we eat carbohydrates again, we will tend to replenish our glycogen instead of building fat. (High-intensity exercise such as fastercise promotes the insulin-induced replenishment of glycogen.)[3] I believe fostering meal swinging

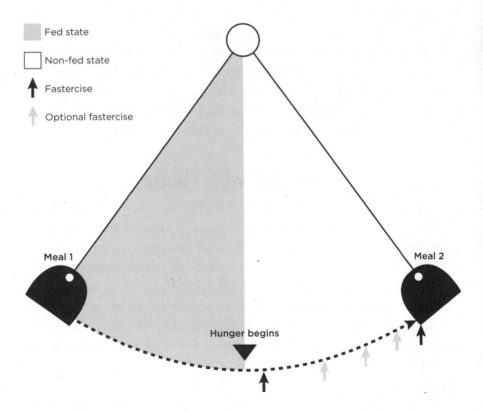

Figure 6.1. *Meal swinging* is my term for getting into a rhythm that allows your body to make a full swing through the fed state and well into the non-fed state before you eat again. Timing fastercise to tamp down your hunger as you enter and swing through the non-fed state allows you to remain comfortable while maintaining optimal hormonal conditions.

with fastercise promotes fat burning when hormone conditions are ideal for fat burning and muscle building when hormone conditions are ideal for muscle building. This enables us to make a great deal of progress with very little effort by working with the body's priorities instead of against them.

Building Momentum

When you are swinging on a swing set, you swing back and forth in two opposite directions. However, even though those directions oppose each other, they also enhance each other because the farther and higher you swing in one direction enhances how far and high you will swing in the other. In a similar way, fat burning and muscle building can oppose and enhance each other when we time our diet and exercise correctly. The hormones that favor fat burning are counterregulatory to the hormones that favor muscle building. However, the non-fed state that promotes fat burning in the moment also increases our sensitivity to insulin that will promote muscle building when we enter the fed state. And the fed state that promotes muscle building in the moment also increases our sensitivity to the counterregulatory hormones that will promote fat burning when we enter the non-fed state. Our muscle building and fat burning can amplify each other, meal by meal. This concept of meal swinging applies for any type of healthy eating plan, including high-carb and low-carb approaches.

I'm sure you've experienced an occasion when you were feeling only a little hungry (or not all), but you decided to eat a little something anyway

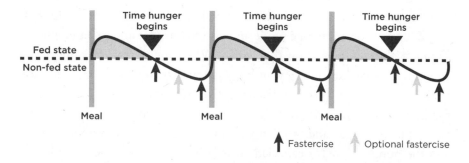

Figure 6.2. When we allow the body to swing widely through the fed state and the non-fed state meal by meal, we amplify the benefits of both to promote vibrant good health.

because you could. And after you ate it, you had a "blah" feeling and felt worse than before you ate it. You might have thought, "What a waste of calories. I wish I hadn't eaten that." So, the next time you get a little hungry, instead of squandering your culinary enjoyment by answering your hunger with food right away, try answering your hunger once or twice with fastercise first. This will heighten your enjoyment when you do eat your next meal. Instead of gaining more fat and enjoying food less, you can burn more fat and enjoy food more.

When we eat as soon as we get hungry, as in figure 6.3, we are not allowing ourselves to swing very far into the non-fed state. This is problematic, because as discussed in previous chapters, when we spend a lot of time in the fed state, we can develop insulin resistance. Being responsive to insulin helps us preserve and build muscle and makes it easier to store fat. But when we overfeed ourselves, we have only small slivers of time in the non-fed state during which our hormones are conducive to fat loss. The swing in figure 6.3 isn't swinging very high, because we're not building much muscle (because of insulin resistance) or burning much fat.

When we underfeed ourselves by not eating enough to satisfy our appetites and not fastercising, as in figure 6.4, then we spend less time in the fed state and become hungry sooner after a meal. We are in the less-fed state, and our metabolism starts to slow. Our bodies become resistant to the counterregulatory hormones, which means we don't access our fuel stores as strongly as we would with more time in the fed state. Ultimately we don't burn as much fat, and hormone levels are conducive to muscle building for only a small sliver of time.

As I explained in chapter 5, fastercise signals our bodies away from storage mode and toward forage mode and also to burn fat and to build muscle. In practical terms, here's how that works. When you start to get hungry and you fastercise, you are letting your body know that you are paying attention to the hunger signal and that you are actively engaged in seeking out more food. Receiving the fastercise signal, your body cooperates by mobilizing stored fuel (glycogen and fat) that your body can burn to power your efforts and the hunger signal is canceled. This is handy because the sensation of hunger would be distracting while you are actively engaged in foraging. At the same time, when you fastercise shortly before you eat, you send a signal about which muscles you want to build. The body gets the message, "These

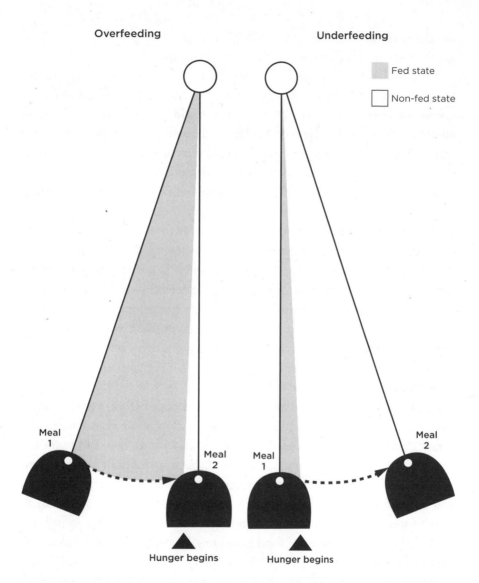

Figure 6.3. Overfeeding increases the time spent in the fed state and decreases the time in the non-fed state. This is like stopping a swing just after it reaches the bottom of the arc, and then trying to push it back the way it came. It fights against the body's opportunity to burn fat and hinders muscle building by promoting insulin resistance.

Figure 6.4. Eating low-calorie meals puts the body in the fed state for only a short time before hunger arises. This fights against the body's opportunity to build muscle and hinders fat burning because it encourages a slowing of metabolism and resistance to counter-regulatory hormones.

are the muscles I used to obtain food." And the body replies, "Then those are the muscles I'll focus on building for next time."

Here is the simple pattern of fastercise and meal timing that promotes forage mode: get hungry, fastercise, eat, repeat. I feel it's important to fastercise as soon as you start to feel hungry. This could be as simple as tightening all your muscles as hard as you can for ten seconds. It's important to pay attention to the hunger signal and never let your hunger go away on its own (it will come and go on its own as it builds); otherwise, your momentum and metabolism will slow. Whenever you notice hunger, it's important to either fastercise or eat lightly to keep your meal-swinging momentum going. After fastercise, your hunger will go away for a time. When it comes back, you can push off hunger again with fastercise if you wish. You can keep pushing off hunger several times if you wish to spend more time in the non-fed state and burn more fat (as in figure 6.1 on page 90). However, if you are in a busy time of your day and find the hunger and fastercising too distracting, you can go ahead and eat lightly to push off your hunger for a time, and resume fastercising later when you get hungry again.

Fastercising *Before* Meals

I also strongly recommend that you fastercise shortly before you eat (within five to ten minutes). This action tells your body that you are a successful forager and that you don't need to store fat but you do need to build the muscles you just used to obtain the food. Fastercising will also suppress ghrelin release, and less food will be needed to satisfy your appetite. When you eat without fastercising right beforehand, you signal the body that you didn't forage and that you are depending on stored food. This signal also tells the body that forageable food is not readily available, leading to a shift into storage mode.

At times, it is permissible to fastercise once when your hunger manifests and then eat lightly right away. In this case, the single bout of fastercise satisfied both requirements. You fastercised when you got hungry *and* you fastercised shortly before you ate. And because you ate lightly, your body won't stay in the fed state too long. These small meal swings are still in compliance with forage mode.

When you fastercise immediately before eating, your body interprets the signal as an indication that you foraged and are not relying on stored food, that there is plenty of forageable food available, and that you can easily obtain more food. And when you eat lightly (just enough to satisfy your appetite), you signal your body that food is so abundant that you can easily meet your needs with foraging. Therefore, you don't need to store a lot of body fat.

When you eat when you're not hungry (I call this overeating), either between meals or at the end of a meal when your appetite is already satisfied, then you are signaling the body that you see this food as a rare opportunity (food is scarce). This signal shifts your body toward storage mode, and can explain the feeling of lethargy that occurs after overeating. When I changed my own eating habits and started eating lightly (only until satisfied), I noticed I felt energetic all day and didn't suffer from lethargy after meals.

When we eat, it's helpful not to eat too little or too much. If we don't eat enough to satisfy our appetite, we may be putting ourselves in the less-fed state, which can slow our metabolic rate and encourage storage mode. On the other hand, if we overeat by continuing to eat after our hunger is satisfied, we may be putting ourselves in the more-fed state, which can encourage storage mode and increase our fat set point. By eating just the right amount of food, we signal the body that foraging is going great and we don't need to shift toward storage mode and that we'd be better off building muscle instead of storing fat. Wow, that's a fine line. How are you going to know whether you are eating too much or too little? You're going to let your appetite do all the calculating and counting of your calories. You're going to pick some healthy foods, as we discuss in the chapter 8, and then you're going to stop eating as soon as your appetite is satisfied.

Practical Tips and Techniques

When you first try fastercise, you may want to ease into it gradually, especially if you haven't been exercising much or if you have health problems. In either of these circumstances, the body's level of mitochondria may be lower than normal. If you are debilitated, it might take you more than a day to fully recover from a standard bout of shivercise. Start small, with just a few seconds of fastercise at a time, and gradually build up from there. By adjusting the amount you eat and the number of times and the intensity with which you fastercise, you can easily work fastercise into any schedule.

Stationary bikes and treadmills often have meters that let us know how much exercise we've done, and pedometers are handy for counting the steps we've taken in a day. Hunger is also a great exercise meter. When we get hungry, we know we've burned enough calories to run out of easily available fuel. When we are feeling hungry and we fastercise, usually our hunger goes away. If it doesn't, then our built-in "exercise meter" is indicating that our glycogen is depleted, or we're dehydrated, or we didn't fastercise hard enough or long enough to mobilize our energy stores. When we answer hunger with tightercise, our appetite can be gone by the time we let our muscles relax. The great thing about using hunger as an exercise meter is that it can save us from wasting too much time or energy exercising unproductively. Exercising too much can deplete glycogen stores and intermediates of the citric acid cycle, which we need to help with burning fat and preserving muscle. Overtraining can also potentially lead to fatigue and injury. We can use hunger as a guide to do just the right amount of exercise at just the right time to serve our purposes. Sweet!

Focusing on Shivercise

Shivercise simulates the small shaking movements of our muscles we do when we shiver in order to use up ATP very quickly and warm our bodies. Shivercise involves very quick contraction and relaxation of opposing muscles in an alternating fashion, just as you would when you shiver naturally.

How fast is fast enough? You may wonder how rapidly you need to move your muscles while shivercising in order to get the benefit. Shivercise essentially boils down to a race between how fast the body can use up ATP and how fast the mitochondrial process can make it. Your goal is to shivercise fast enough to leave your mitochondrial process in the dust. As you pull ahead in the race, your breathing will increase and your muscles may start to burn. As soon as you build up a big enough lead, your need to breathe will increase to the point that you'll be able to gasp for and catch a deep breath. If you don't get winded enough to catch a deep breath within two minutes or so, you probably aren't moving fast enough.

As you continue to shivercise every day, you may find that it becomes harder and harder to get winded enough to catch that deep breath. This is a good sign that you're developing more and more fat-burning machinery. (Your mitochondria are getting larger and more numerous.) Another good sign that you are fastercising fast enough is that your hunger disappears within about a minute. Usually, the more intensely you shivercise, the longer your hunger will stay away (provided you have enough glycogen and hydration).

Stay in the lead, and breathe. Think of shivercise as a race between you and a slower competitor. Your goal is to establish as big a lead as quickly as possible. If you go fast enough, your lead increases, but if you slow down, your lead doesn't build much, if at all. If you slow down too much, your lead may even decrease. When we shivercise, we start breathing faster. Breathing clears CO_2 and lowers acid levels in the body's tissues, and acid buildup is what stimulates our ability to catch a deep breath.

Just get started. Although leaping into the lead of the imaginary race has benefits, there's no rule that says you have to burst into shivercise at 100 percent. You can ease into it at any pace you like. The key is to get started. Why not start easy for a few seconds? Even when I don't particularly feel like fastercising, it's easy to just get started. And then I discover that it's easy to speed up from there, and before I know it, I'm done.

This is the moment. Perhaps the greatest benefit of fastercise is that so much benefit can be generated by a matter of seconds' worth of effort. We can often accomplish more in strategically placed moments of high-intensity effort than we might otherwise accomplish in hours of poorly timed lower-intensity effort. Even though it's only a matter of seconds, we still need to make the effort. I like to remind myself that the effort I expend while shivercising will increase my ability to burn fat for the rest of today and beyond. How badly do you want your body to burn fat and build muscle, to transform your body, to get in the best shape of your life? This is the moment (along with other moments like it) that you can make that happen. The faster you shivercise, the more powerfully you signal your body. There are lots of ways to motivate yourself to gradually amp up your speed while shivercising. Put on some high-energy music! Or, focus on going a little faster every few seconds.

Even though it's hard to maintain peak intensity for very long, there are many verified stories of people briefly performing near-superhuman feats of strength for a few seconds in an emergency situation. Focus your efforts of increasing intensity in waves of two to three seconds, like climbing stairs. Sometimes, I repeat the word *faster* in my mind every few seconds as I level up. (Speaking of superhuman, as you continue to fastercise regularly you might be astonished by the profound changes you see in the form and function of your body, almost as if you actually are developing super-powers.) Racing with that person in the mirror is another way to focus. Sometimes I build up my speed close to the "summit" of my shivercise and then when I'm ready, I "make a run for it." That's when I focus all of my energy for the last few seconds to race ahead of my mitochondrial process as far as I can as fast as I can.

Go faster. A little extra intensity can make a huge difference in a shivercise session. Imagine you're racing someone named Mito whose top speed is five miles per hour. If you run five miles per hour, too, you'll stay neck and neck. If you speed up to six miles per hour, then you begin to pull ahead of Mito. But if you go *just a little* faster—seven miles per hour—you could increase your lead twice as fast! If you speed up a little more to eight miles per hour, suddenly your lead is building *three times faster*! In a similar way, if we move just a *little* faster while shivercising, we can overwhelm the mitochondrial process much more quickly. Going *much* faster for a matter of seconds can often accomplish more than going a little faster for much longer. The key is the top speed, or intensity.

Try a second round. If we can't quite accomplish catching a deep breath a few seconds after a bout of shivercise, we can try a second bout. The less time between bouts the better, because it gives the body less time to clear the acid buildup.

Add some variety. Shivercise can take many forms. You can run a sprint, or sprint for several seconds on a stationary bike. You can lie down on a couch or bed and shake your muscles with small movements. For example, you can flutter kick your legs as though you're swimming. You can do the same thing hanging from a chin-up bar or sitting on a chair, recliner, desk, or countertop. The point here is that you don't need a lot of resistance, or even any resistance. The inertia of the rapidly alternating movements provides resistance in itself. But whether you use any resistance or not, speed is still the key. The faster you go the better it works. Shivercise is about moving so fast that you have to stop, not moving moderately so that you can stay in motion for many minutes. You can shivercise in a swimming pool. You can also dance to fast music, with your arms raised or lowered, twisting your hips left and right to exercise your core muscles.

One convenient approach is simply to run in place, moving your arms and legs only a few inches back and forth, lifting only your heels off the ground. It's easier to move faster when you make small movements. When I first started shivercising I used this approach and I still do. At first, I would get winded and catch a deep breath in twenty seconds or so. But as I continued shivercising every day, it took longer and longer for me to catch a deep breath. Now it sometimes takes me as long as two and one-half minutes. It's harder for me to go fast enough to get winded, I believe, because of the mitochondria I have built up. You don't have to keep increasing the amount of time you shivercise in order to maintain your improvement; however, I like to maximize the benefit so I make sure I shivercise long enough to catch a deep breath at least once every day.

I find that I can get winded faster by making my shivercise movements smaller and easier. I shake or shiver intensely the way I would if I were cold (smaller movements than running in place). When we shiver, we shake our muscles back and forth in opposing directions. I believe it's the speed at which we change directions that burns the most energy. Just as you can warm your hands faster with smaller movements, you will likely find it easier to get winded faster by shivercising with smaller movements. I can catch a deep breath more than twice as fast by shivering than by running in place, which is astonishing. This has demonstrated to me why we automatically shiver when we're cold instead of automatically running in place. Better results with less effort and more comfort. Perfect.*

Focusing on Tightercise

Unlike shivercise, tightercise is all about holding our muscles fast (still) and tightening them as hard as we can. Here are some tips for getting the most out of tightercise.

It's just stretching. Sometimes we don't feel like exercising, and it can be impossible to find the motivation to do a workout. One of the beauties

---

* You can find video demonstrations of shivercise variations at www.fastercise.com.

of tightercise is that it requires only a few moments of effort tightening our muscles hard, as we do when we stretch in the morning, and you know how good that can feel. Sometimes I don't feel like tightercising; however, I remind myself that it's not a workout, it's just stretching! I start tightening my muscles with the idea of re-creating the enjoyable sensation of a good stretch. Before I know it, I'm done and feeling energized.

Snack on storage. When you start to get hungry and you tightercise, you signal your body to mobilize stored energy and your body is immediately satisfied. You just directed your body to snack on its storage. If you want, you can push off your hunger with tightercise again when your hunger returns. It just takes a matter of seconds every so often to keep your body satisfied snacking on storage. Tightercise followed by a deep breath or a yawn can cancel hunger, prepare us for physical activity, or prepare our bodies for the fasting associated with sleep.

Nurture your tightercise habit. Our bodies respond beautifully to training. Transitioning from a dependence on food to a dependence on stored fuel takes some conscious effort. When we're dependent on food, eating can become so habitual that it can get to the point of being almost unconscious. We can be chewing and swallowing food almost before we know it, and keep eating even after our appetite is satisfied. In a similar way, once we get used to canceling hunger with tightercise, we may spontaneously start feeling an urge to tighten our muscles hard as our bodies start to enter the non-fed state. I have caught myself tightercising for a few seconds without having consciously chosen to do so. I've even had the experience of realizing in the evening that I barely noticed my hunger for most of the day—I attribute that to habitual tightercise responses that I wasn't even aware of in the moment.

> "I used fastercise mostly in the evenings because I am a snacker. It was effective when everyone around me was snacking. Fastercise gave me something else to focus on and it abated my cravings."
>
> —LISA H.

Go for full contraction. Charles Atlas, a famous bodybuilder in the 1920s, popularized isometric exercises for muscle building. Tightercise is potentially more effective when a muscle is fully contracted, or balled up, because of the way in which muscle contraction occurs. (I explain the physiology of this in chapter 7.) The goal with tightercise is to reach peak intensity by contracting your muscles as hard as you can, anywhere from a few seconds to about a minute. You may worry that your muscles will cramp if you do that. However, if you are well hydrated and ingesting enough electrolytes, cramps should not be a problem.

Choose your target muscles. It can be difficult to figure out how to target a certain part of a certain muscle when weightlifting or using exercise machines. All the angles and forces can be hard to sort out, even for a physics major. However, with tightercise it is easy. Simply put your finger on the muscle you want to develop. Then, move your body in whatever way you need to in order to tighten that muscle into a hard "ball."

Reach for the cramp. Have you ever strained to reach an object and felt one of your muscles starting to cramp? You can use the same approach with tightercise. Try tightening your muscles by pretending to reach for something just out of reach as hard as you can until you feel as if they are about to cramp, and then hold it for a second or two. To make absolutely sure you are activating the correct portion of the correct muscle, you can use the finger trick mentioned above and place your finger on the very muscle portion you want to activate to ensure that the "about to cramp" sensation correlates with where your finger is. The muscle-building stimulation from tightercise increases with intensity.

Start with one, then add more. You can get a lot of tightercise done in a matter of seconds by tightening many muscle groups at the same time. To do this, choose a comfortable position, whether that's sitting, standing, or lying down, and focus on tightening up just one muscle group. Then add another and another until you are tightening many muscle groups at once. For example, keeping your muscles tight, try adding groups in this order: upper arms + forearms + hands + deltoids + chest + back + abdomen + buttocks + front of thighs + back of thighs + front of legs + back of legs. You won't be moving, because you're tightening all the muscles against themselves. Keep shifting your mental focus to be sure muscles groups remain tight. You can tighten muscles this way for one minute with nearly no chance of injury.

Go with what feels good. When we wake up in the morning, the way we do our morning stretch varies from day to day. We might flex our arms, or we might extend them. Another day we might flex one arm and extend the other. It depends on what feels good in the moment. In addition, we sometimes adjust our position to tighten certain muscles while letting others relax. In a similar way, we can do some "migrating" tightercise. The chest, back, and shoulders have groups of muscle fibers that fan out in different directions, the way fingers fan out on a hand. We can sweep our bodies through different movements to sweep through a fan of muscle fibers. By reaching for the cramp while isolating small groups of muscle fibers for a few seconds at a time as we sweep through those fibers, we can easily exercise our muscles more effectively in just a minute or so than we might otherwise in an hour at the gym lifting weights, and much more naturally and proportionately as well.

Techniques to try. There are innumerable ways to tightercise, but here are some ideas. Lie facedown on the floor or on a bed and raise your arms and legs like a skydiver and slowly sweep your arms and legs as if doing jumping jacks or making an angel in the snow. In the process, tighten your muscles as hard as you can. You can try reaching for the cramp by trying to reach for the ceiling while you are sweeping through these movements. Point your toes so you can contract your calves, and then flex your feet tightly to contract your shin muscles and your quadriceps. Try balling up your fists and flexing your forearms to the max, then alternate by extending your fingers and bending your hands in the opposite direction.

Sit on a table or countertop supporting your weight on your hands as you lift yourself off the countertop and draw your legs up as high as you can. Then, focus on tightening hard the back of your upper arms, some of your chest muscles and back muscles, and perhaps some of your shoulder muscles. Tighten your abdominal muscles, and then straighten your legs and tighten your thigh muscles and lower leg muscles. Stand up and turn your hips, shoulders, and head as far as you comfortably can toward the left and then tighten all the muscles involved in the twist. Release and try the same twist but looking to the right.

Chin-up bar moves. While hanging from a chin-up bar, curl your legs up at the knee and tighten hard the back of the legs. Straighten your legs and draw them up in front as high as you can to tighten your abdominals and thighs. Extend one leg forward and one leg backward in a hurdling pose, tighten for a few seconds, and then switch legs and tighten again.

We can also simply tighten any muscles we feel like tightening the way we do in the morning, shifting from position to position as we go. And we can tightercise anytime we feel like stretching because our bodies are already inclined toward it. We can consciously amplify our automatic stretching inclinations and add to them as we wish.

Ball up your muscles. Standard exercise positions such as push-up or plank position can work well for tightercise, too. In push-up position, try tightening very hard some of your chest, arm, forearm, abdominal, and leg muscles. Keep in mind that it may be most beneficial to tighten muscles that are in the balled-up position. For example, when you are in plank position the triceps muscles can be in the shortened, balled-up position but the biceps on the front of your arms are not. That's why it works well to sweep your body through a variety of positions so that you can ball up (shorten) different muscle groups in turn and reach for the cramp.* As you can see, there are many ways to tightercise but it just boils down to tightening the muscles you want to develop as hard as you can for a matter of seconds.

Which One When?

I believe that both shivercise and tightercise are very useful in helping us lose fat and gain muscle. They each have their benefits and I wouldn't want to do without either of them. To get the most benefit from fastercise, it's helpful to understand the similarities and differences between tightercise and shivercise. Both signal the body to burn fat and preserve muscle. Both eliminate hunger. However, shivercise burns ATP more quickly than tightercise, while tightercise may build muscle size and strength more than shivercise. Because shivercise burns ATP more quickly, it can send a very powerful signal to the body that it's time to mobilize stored energy in order to eliminate hunger and to increase the number and size of our mitochondria. When we first begin to fastercise we may need a signal this

---

* For videos on tightercise, you can visit the fastercise website at www.fastercise.com.

powerful to eliminate our hunger. However, because shivercise burns ATP more quickly, we tend to burn up our reserves faster and may not be able to put off hunger for long. In addition, when the hunger does come back after a bout or two of shivercise, it tends to come back more intensely than with tightercise.

Tightercise can be done before or after we've eaten. For muscle building, it's best for us to provide amino acids to the body a little before or after tightercise. As we begin to fastercise regularly we notice that we don't get winded as easily. This indicates that our capacity to make ATP is increasing. We may find that with this increased capacity, tightercise eliminates hunger quite easily. At the same time, we might be able to put off our next meal longer and more comfortably with tightercise than shivercise. And we still get the muscle-building benefits of tightercise, particularly when we supply the body with amino acids. In this case, we may prefer using tightercise to answer hunger instead of shivercise. The key to all of this is for us to pay attention to the signals the body is sending us and the signals that we are sending the body to see what works best for us now and as we progress. Both tightercise and shivercise are extremely valuable forms of fastercise. We may find that we prefer one or the other at different times for different reasons.

When I first started fastercising, I found that shivercise did a better job of curbing my hunger. But as time went by (and, I presume, my body developed more mitochondria), I found that tightercise was just as effective at making my hunger go away. Over time, I've found that tightercise tends to build more muscle than shivercise. What works for me, though, may not necessarily work for you. I recommend doing shivercise at least once a day in order to promote mitochondrial growth. Beyond that, you can cancel your hunger with either tightercise or shivercise depending on what works best for you. Here are some other thoughts to help you decide which form of fastercise to do when.

Try tightercising first. Sometimes it's handy to tightercise right before you shivercise. That way, you get in a little resistance training and build up acid in your muscles, which provides a head start on that deep breath you want to achieve with shivercise.

Shivercise before a feast. One reason to shivercise is to build up our fat-burning machinery by increasing the size and number of our mitochondria.

Shivercise is great to do when you're hungry, but there are also other times shivercise can be helpful even when you're not hungry. Shivercise will make you more alert, so it's helpful to do it first thing in the morning, and also thirty to sixty minutes after lunch to help fight off the afternoon doldrums. (Eating lightly also helps prevent the doldrums.) Shivercise can also help you prepare your body for consuming an extra-large meal. If you'll be attending a gathering and you know there will be special and lavish food, it's best not to walk into the situation with full glycogen stores. If you do, the abundant excess carbs from the meal will go to build up your fat stores. So it's smart to empty your glycogen stores as much as you can before feasting. Shivercise is very effective for using up ATP quickly. (I explain the science behind this in chapter 7.) To prepare for an evening feast, shivercise once or twice in the afternoon, even when you don't feel hungry, and again when you begin getting hungry, then shivercise once or twice more (and don't eat yet!). You'll know your glycogen stores are almost empty when you can't hold off hunger more than thirty minutes with shivercise.

Do both to beat stubborn fat. A combination of shivercise and tightercise may be helpful in addressing stubborn fat. Stubborn fat is the fat that comes off our body last and it responds differently to epinephrine than visceral fat does. Epinephrine may actually inhibit the breakdown of stubborn fat to a degree, which may be what makes it stubborn. However, one study indicated that as we get more fit, the inhibition of stubborn fat breakdown by epinephrine may be reduced.[4] For me, meal swinging has been the most effective approach I have ever found for losing my stubborn fat. It's hard for me to imagine how anyone could lose stubborn fat without consistently sending forage signals to their body by doing fastercise. It's possible that people who have jobs that require lots of daily physical activity or who exercise all the time might manage stubborn fat without fastercise, but for the rest of us, I recommend a combination of both fastercise techniques to fight stubborn fat.

Before an athletic event. Doing some shivercise or tightercise before athletic competition is like doing a warm-up. On the positive side, shivercise and tightercise increase levels of various hormones, as well as blood flow, respiration, ketones, and mental focus. But there is a negative side, too. Fastercising can deplete on-hand ATP and glycogen, for example. So it seems that a little fastercise might be helpful but a lot might be too much. How much is enough and how much is too much will vary on a case-by-case basis. However, we do have a hint from the animal kingdom. Some videos of predators hunting prey show that the predator stretches (tightercises) for a few seconds, and then yawns shortly before charging after their prey. Were they sleepy or bored? I don't think so. I think they were doing just enough tightercise to prepare themselves for the burst of energy and strength needed to capture their prey.

Before eating. As explained earlier in this chapter, the more we remember to tightercise shortly before we eat a meal, the stronger and bigger our muscles become. Such is the effect of resistance exercise like tightercise.[5] Tightercise can be beneficial whether done in the non-fed state or the fed state. In the non-fed state, it can curb hunger and help preserve muscle, and stimulate the muscle to grow when fed. Tightercise sensitizes muscle synthesis to amino acids for twenty-four to forty-eight hours.[6] Sometimes, after a meal, we can feel revived and we can feel our strength coming back, just as when horses are "feeling their oats" after being fed and are feeling frisky or animated. This can be a wonderful time to tightercise because our bodies actually want to exert themselves at such a time, and amino acids may be plentiful.[7]

More Benefits of Fastercise

Discomfort and inconvenience can be huge obstacles to losing fat. It can be difficult to muster the motivation to spend a lot of time doing an activity we don't enjoy, particularly when we don't see the results we're looking for. Fortunately, people find that high-intensity exercise like fastercise carries lower perceived exertion than steady-state exercise and is associated with greater fat loss.[8] And the following ancillary benefits of fastercise may prove as appealing and worthwhile for you over the long term as does its effectiveness in helping you get in shape.

Reducing Stress

Fastercise can be helpful as an outlet to stress. When we are stressed, our bodies build up hormones in preparation to fight or flee. Only, in the modern world, we usually don't physically fight or flee—we just stew in our stress. Fastercise simulates fleeing and may provide a constructive outlet for pent-up stress. Think of how good it feels to stretch first thing in the morning or after sitting for a long period during the day. A short bout of fastercise can be especially helpful in letting go of stress if you also focus on positive thoughts and follow the fastercise with a moment of conscious relaxation.

Feeling and Looking More Fit

Have you ever noticed when people gain muscle and lose fat, their skin also tends to look better? As a person gains fat, their growth hormone levels can drop and their skin may sag, but with fat loss, GH levels increase.[9] The production of GH is modulated by many factors, including age. The structure and appearance of the skin directly reflect the level of GH production that occurs in the different phases of life.[10] GH levels decrease by 14 percent every decade, starting at age thirty.[11] This decrease accounts for much of the sagging and wrinkling of skin over time. Because high-intensity exercise such as fastercise naturally increases levels of GH, it has the effect of tightening and repairing the skin, which can result in a healthier and more youthful appearance.

You've probably noticed that when you use certain muscles in your body more, they tend to get bigger. And muscles you don't use tend to get smaller. Use it or lose it, and fastercise is a great way to direct the body to preserve muscle. When we shivercise and tightercise, we can focus on using the muscles we want to preserve. And we know that intense exercise works the fast-twitch white muscle fibers, fostering bigger muscles, like those of sprinters or gymnasts.

Have you noticed that as you've gotten older you can't move around as easily? GH, increased by fastercise, can help repair your muscles, skin, bones, and joints. Fastercise can help you lose visceral and subcutaneous fat, build muscle, and improve flexibility. Imagine being able to fit into clothes more easily and put them on more easily. Can you tie your left shoe while standing balanced on your right foot? Imagine being able to do things you haven't been able to do in years.

Improved Respiratory Health

High-intensity exercise like fastercise enhances both aerobic (mitochondrial) and anaerobic (cytosolic) fitness levels.[12] This enables us to exercise more before getting winded and recover faster when we do get winded.

After I began fastercising, I noticed I wouldn't get winded as easily by physical activity. I tested myself by timing how long I could hold my breath. To my amazement, after only a few months of fastercising, I was able to hold my breath 50 percent longer than I had ever been able to! Plus, when I got winded, I was able to recover normal breathing much more quickly; it took only a few deep breaths. After all, the cytosolic and mitochondrial processes that fastercise strengthens are collectively known as cellular *respiration*. (Just sayin'.) I believe that fastercise will be helpful for anyone—from a sedentary person to an accomplished athlete—who wants to develop better respiratory performance.

Higher Productivity

Fastercise can increase our energy and sharpen our thinking, which makes sense considering we would need energy and sharp thinking to escape dangerous situations and to obtain food. A bout of tightercise can be fantastic for very short breaks at work. We can quickly cycle through the muscles we want to hit, a few seconds each, and be done within a minute. This can be extremely invigorating and can help us focus and get our work done well. Shivercise also helps boost the counterregulatory hormones and productivity.

Why take a substitute when we can easily have the real thing? Many people try to improve their alertness and productivity by starting their workday with a fastercise substitute called coffee. The caffeine in coffee is known as a sympathomimetic stimulant.[13] *Sympatho-* refers to the sympathetic nervous system responsible for fight or flight. The quintessential sympathetic hormones are epinephrine and norepinephrine. *Mimetic* means "imitating." Thus, the hormones that caffeine mimics are epinephrine and norepinephrine, which are hormones increased by fastercise. The cream and sugar people put in their coffee mimic the fat and glucose fastercise releases from adipose tissue and glycogen respectively. One big difference, though, is that when you burn your own energy stores you get thinner, but when you burn fuel you're putting in your mouth there's always the chance you could overdo it and gain fat. Another big difference is that fastercise provides predigested and energized glucose in the form of lactate for sharp thinking that coffee doesn't provide. Fastercise has a faster and stronger effect than coffee (judging by heart

Fastercise Break Rooms?

Breaks that involve movement, nature, and full detachment are especially helpful in reducing work errors and boosting mood.[14] The top companies in the United States have health and wellness programs to help their staffs be more productive, creative, dynamic, and successful as teams. In addition, wellness programs can provide lower stress, fewer sick days, lower insurance costs, and improved family life. Many of these companies provide state-of-the-art fitness centers. Although it's easy to inconspicuously do tightercise while sitting at a desk, it would be ideal to have quick, easy access to spaces optimized for incorporating tightercise and shivercise into our workday. Just as companies provide easily accessible restrooms, I can picture more progressive companies setting up easily accessible spaces, preferably outdoors in nature, or in a room with a natural motif and the option of voice-activated music, where employees could go for a quick dose of fastercise to recharge and refresh.

rate, breathing, and body heat) and is less time-consuming, less expensive, and also provides the health benefits of exercise (fat burning, muscle building, mitochondria building, endurance building, strength building, and more). Perfect. Appetite-suppressant pills are also sympathomimetic stimulants. Do you see the relationship? Fastercise suppresses appetite, too. During the day, hunger pangs can be very distracting and decrease our productivity. It's so nice that we can immediately cancel those pangs with fastercise.

When I give presentations at conferences, one reason I like to lead attendees in a minute of shivercise is that epinephrine not only boosts energy and mental focus but it also stimulates nerve growth factor, which aids in learning. Fastercising before starting any mentally challenging activity can save you from the trap of losing focus and having to read the same sentence or paragraph two or three times over.

The time-management method known as the Pomodoro Technique, created by time-management consultant Francesco Cirillo, involves setting a timer and working for twenty-five minutes, then taking a five-minute break. This technique promotes focus while working, as well as recovery during breaks. A five-minute break is a perfect amount of time to assess whether we're hungry or tired and then do a minute of fastercise if needed. This can give us plenty of energy and focus to get back to highly productive work.

Reducing Chronic Inflammation

Oxidation means losing electrons such as to oxygen. For example, oxygen removes electrons from iron to make it rust. Hydrogen peroxide works as an antiseptic by removing electrons from the cell walls of bacteria, destroying them. There are also many oxygen-based household cleaners available. In the body, oxygen and high-energy electrons extracted via the breakdown of nutrients combine via the mitochondrial process to make ATP. Oxidation is super powerful and super helpful, but it can also be destructive. The contained burning of fuel powers much of our transportation, but the US Federal Aviation Administration requires big airports to maintain fire departments. In a similar way, mitochondria provide containment for the enormously powerful mitochondrial process. When oxidation occurs in the cell in places where it shouldn't, it is like a damaging fire breaking out. This is called oxidative stress or oxidative damage. When such a fire breaks out, the body sends out first responders, much like firefighters, to address the problem. This process is called the inflammatory response, or inflammation. Poorly controlled inflammation is at the heart of most chronic diseases. Imagine a plane catches on fire at an airport—soon there are fire trucks everywhere, interfering with runway operations. And for some reason, the fire trucks remain there for days, getting in the way. The airport can't function properly, and neither can your body when inflammation interferes with the proper function of organs and systems. Fortunately, high-intensity exercise reduces oxidative stress and inflammation.[15] Just as firefighters gain experience with practice and fortify their processes and equipment, so, too, can exercise stimulate the body to refurbish its processes and equipment (that is, cells).

But exercise alone isn't enough to combat inflammation; other aspects of self-care play a role, too. Just as firefighters need adequate funding for protective clothing and equipment, we also need to provide our bodies with the protein, oils, and nutrients they need to maintain strong membranes and cells. If our bodies have the resources and ability to do the repairs and refurbishing, then the stimulus of high-intensity exercise like fastercise can be purifying and strengthening. If they don't, due to poor nutrition or disease, then it's conceivable that high-intensity exercise can increase oxidative stress. Because we engage in fastercise in small doses, the risks are also small and we can readily monitor whether fastercising helps with an inflammatory condition or not. Of course, this caution applies to all forms of exercise. Although exercise is crucial for optimal health (decreased inflammation), the barrage of advertising for products to provide athletic pain relief is ready evidence that exercise can also contribute to increased inflammation (in the form of pain, swelling, redness, and heat in joints, muscles, and tendons).

Beating a Chill

Another handy use of fastercise is simply to warm up if you're feeling chilled. Long before our bodies cool down enough to trigger shivering involuntarily, we can shiver and shudder voluntarily (shivercise). This motion releases a tremendous amount of heat that can warm the body much faster than drinking a hot beverage, and without scalding our mouths. We can warm our bodies by burning our fat stores rather than adding to them. Have you ever been in a room where the thermostat is set so low you feel like you're freezing to death? Shivercise and tightercise can warm you up quite quickly. If it would be awkward for us to shivercise where we are, we can tightercise instead. We can tighten all our muscles really hard wherever we are sitting or standing. No matter where we are or what we're doing, we can almost always tightercise to warm up.

Fastercise for Those with Limitations

Some fitness experts are opposed to the idea of fast, high-intensity exercise, especially for people who are out of shape, for fear of joint and tendon injury. However, I feel that fastercise can be very easy on the joints and tendons and can be safely and easily and comfortably performed by most people who are out of shape. Fastercise will help them get in shape and strengthen their joints and tendons and muscles. After all, most people can safely shiver when they're cold and stretch when they wake up.

Fastercise is very easy to do anytime during the day, every day. This is handy for easily dosing our exercise throughout the day rather than driving to a gym and doing it all at once. For example, a bout of exercise increases the secretion of GH within ten minutes of exercise and the GH tends to go back down again over the next hour.[16] Consequently, repeated exercise every hour or two is handy for burning fat. We can keep our GH levels up by answering hunger with fastercise. In addition, as we get older, GH tends to go down (often referred to as somatopause). Fastercise can be an easy way to get the regular exercise that is very beneficial to us as we age.[17]

Once, I showed doctors a simple form of shivercise that involves vigorously shaking two partially filled sixteen-ounce disposable water bottles. One doctor told me later how he's been using the concept in his patients with great results: "I have patients who are able to shake water bottles when they aren't able to do any other form of exercise because of bad knees, bad backs, or other problems."

Tightercise is a great form of exercise that almost anyone can easily do. Even people with disabilities and joint problems can tightercise. For example, I have a damaged rotator cuff in my right shoulder. I damaged it weightlifting decades ago and from time to time since. If I'm not careful, I can easily damage it further if I work with weights. On the other hand, I can tightercise and build all the muscles around my shoulder including deltoid, chest, and back without further damaging my rotator cuff.

We can tightercise anywhere at almost any time. In addition, tightercise is quite safe. We don't have to worry about spotters, or fitness center accidents or injuries. Even sitting in a wheelchair or at a desk at work, it's possible to tighten groups of muscles hard and then release them. And we can tighten many muscles at the same time, which is extremely efficient. Tightercise can be especially helpful for the elderly. They can easily tighten their muscles when they wake up and also when their hunger reminds them. This can build up their muscles to strengthen and improve their mobility, and maintaining mobility has been shown to reduce risk of death from any cause.[18]

CHAPTER 7

The Physiology of Fastercise

I magine that your house is right at the foot of a big hill. Every time there's a heavy rain, a mudslide happens and you have to spend hours raking and shoveling your yard to clean up the mess. A better solution might be to build an embankment to prevent the mud from coming into your yard in the first place! Trying to exercise your way out of bad eating habits is like cleaning up the mud every time it rains, because it takes so much exercise time to burn a significant amount of calories. Fortunately, though, you can fastercise your way into healthy eating habits, which is like building that embankment to protect your body from excess fat accumulation. This chapter dives into the physiology of exercise and fastercise to explain how this works at the cellular level.

We can think of fastercise as exercising not only our muscles but also our hormones, hormone sensitivity, cytosolic pathways, and mitochondria. In fact, our real purpose in exercising our muscles is to exercise our metabolisms. Nevertheless, focusing on the type and timing of exercise to influence hormone levels, pathway activity, and fuel sources can generate amazing results in the strength and appearance of our muscles with a remarkably small investment of time and effort. Timing may not be everything, but it can make a huge difference.

Let's start by making a distinction between low-intensity exercise and high-intensity exercise. Cycling or running on a treadmill are good examples of low-intensity exercise, which I also refer to as steady-state or continuous exercise. They require effort, and you may find yourself sweating and breathing fast, but you can keep up that effort continuously for an extended period of time (more than ten minutes), and without having to take deep, gasping breaths. High-intensity exercise is so intense that you won't be able to sustain the effort for more than a few minutes, and it may require deep breathing.

My understanding of the physiology of fastercise is based on research on the effects of high-intensity exercise on the body. High-intensity exercise has become very popular recently and is discussed in many books and exercise programs. Fastercise is similar to these programs in that it involves intense muscle activity done in short bursts. Fastercise is unique, however, because it is the first exercise program to time exercise and meals according to appetite in order to signal hormonal pathways to burn stored fat while preserving and building muscle.

Exercise causes immediate physiological changes as well as adaptations in our bodies (metabolic, cardiovascular, respiratory, and neurological) over time. The adaptations include changes in the concentrations of the enzymes that catalyze the energy-producing pathways, the number and size of our mitochondria, our fuel storage compartments, and how well we access and burn our stored fuel. Exercise can also change our body composition, strength, and appearance.

The more out of shape we are, the more our bodies respond to exercise. For example, epinephrine levels and growth hormone (GH) levels increase (up to six times more for GH) in response to exercise in untrained people compared to trained people.[1] That's great news for those of us who haven't been exercising very much and would like to improve our fitness.

Diet and Fastercize to Maximize TEE

As noted earlier in this book, excess body fat actually decreases fat burning.[2] In other words, when the muscle/fat balance is leaning toward fat, the body burns less fat, which increases the tendency to gain even more fat. Fortunately, exercise can move the muscle/fat balance more toward muscle. A 1988 study involving forty obese women demonstrated that exercise helped participants lose fat while gaining muscle.[3] Exercise can be particularly beneficial for people with high body-fat content because exercise preserves muscle mass and targets the visceral fat (around the internal organs), which is considered more harmful to health than subcutaneous fat (under the skin).[4] Exercising while hungry also increases the body's ability to burn fat.[5] When we are in better shape we tend to burn more fat as compared to carbohydrates during exercise, which makes it easier to stay leaner.[6]

Many people assume that it's necessary to exercise for long periods of time and burn lots of calories in order to lose fat. But our caloric expenditure does *not* increase in proportion to the amount that we exercise. Research by evolutionary anthropologist Herman Pontzer and his colleagues shows that when people exercise above a moderate activity level, their energy expenditure plateaus rather than continuing to significantly increase with increased activity.[7] This doesn't seem logical. After all, it takes energy to exercise, so the more we exercise, the more energy we must be expending. However, consider this key point: The energy we spend on exercise is not our *total* energy expenditure. We also expend energy on other body functions, such as maintaining our tissues, making hormones, regulating our immune system, and typical activities of daily living. It's natural to assume that the amount of energy we spend on these functions remains constant, irrespective of exercise activity. However, Pontzer's research demonstrates that as activity level tends to increase, the amount of energy expended in other areas tends to *decrease*. In other words, the body maintains total energy expenditure (TEE) within a fairly narrow range. This may help explain why people who exercise more tend to experience less chronic inflammation and less autoimmune disease. Because they are spending more energy on exercise, their immune systems may dial back, reducing the tendency to expend energy attacking their bodies' own tissues.

Dr. Pontzer also observed that the TEE of mammals living in the *wild* are similar to mammals in *captivity*. In addition, Dr. Pontzer found that the TEE of people who live as hunter-gatherers and move many miles a day is very similar to that of sedentary office workers. These observations can have some surprising ramifications. For example, the TEE of people confined to bed rest was higher than would have been expected based solely on their activity levels. And the TEE of people who engaged in vigorous activity was lower than would have been expected. Dr. Pontzer and colleagues found that people with greater body-fat percentage actually exhibited a little *more* TEE across all activity levels than those with lower body-fat percentage. This supports the discussion in chapter 5 of the body's survival balance: Our bodies want to maintain their ability to forage and fight.

Dr. Pontzer and colleagues hypothesized that the balance between food availability and physical activity may be a determining factor in maximizing TEE. When we have fuel available and we exercise regularly, our metabolic rate increases. I believe that short bouts (approximately one minute) of exercise also promote energy expenditure. And thus, answering hunger with one minute of fastercise between meals may be more effective at promoting a favorable survival balance (muscle/fat balance) and TEE than running long distances.

As discussed in chapter 5, my theory is that when the body is exercising heavily, it turns down energy expenditures on nonexercise bodily functions to reduce its chances of starving. And by resuming energy expenditures in other areas during periods of low activity, the body may be trying to refurbish itself and to burn excess calories (to keep body weight down) so that it can better hunt, forage, fight, and escape when necessary.

Experts often recommend that we should exercise at least an hour a day if we want to lose weight.[8] However, in general, we can get the same benefit by exercising more intensely for a shorter period of time.[9] The question is, how little time spent going all out is required to provide benefit? The good news is that research now shows that one-minute bouts of high-intensity exercise increase our mitochondria, our capacity to burn glucose and fat, our insulin sensitivity, and our overall health and cardiovascular fitness.[10] In one study, subjects performed ten one-minute bouts of high-intensity exercise each separated by one-minute rests and found it enjoyable.[11] A one-minute bout of high-intensity exercise is a very apt description of shivercise!

Research also shows that *exercise alone* often generates modest weight loss but that *diet plus exercise* tends to generate the best maintenance of weight loss.[12] Thus, one way to think of fastercise is not so much as a technique to burn more calories to lose fat but as a strategic trigger that signals the body to store less fat. And at the same time, fastercise also counters the negative effects of food deprivation and helps us feel comfortable and energetic while consuming fewer calories and burning more of our stored glycogen and fat instead.

The increased metabolic demand of the exercising muscle is the main driving force for all physiological regulatory processes in the muscle.[13] High-intensity exercise can lower appetite immediately.[14] Not only that, but one study showed that a twelve-week program of regular high-intensity exercise resulted in a clinically meaningful decrease in appetite-directed food intake even forty-eight hours after the last bout of exercise.[15] This shows that exercise is effective at controlling intake as well as burning calories. Considering it takes about eight hours of jogging to burn the calories in one pound of fat, it may be that controlling intake is a much more effective use of exercise than burning calories.

Exercise Effects on Hormones

Exercise has profound effects on hormone levels and sensitivity, but I focus on only a few of those effects in this chapter. Exercising muscle certainly requires access to fuel, so it comes as no surprise that exercise increases the levels of counterregulatory hormones (epinephrine, growth hormone, glucagon, and cortisol) that promote mobilization of fuel from glycogen and fat stores. Both resistance and aerobic exercise increase levels of growth hormone, epinephrine, and glucagon. What's surprising is that we can generate those hormonal responses by fastercising for only one minute, when an hour of low-intensity exercise does not produce the same degree of response.

Exercise also increases levels of endorphins, which are peptides (chains of two to ten amino acids) produced in the brain and nervous system that activate the body's opiate receptors, reducing pain. Endorphins inhibit a hormone called somatomedin, thereby increasing the release of growth hormone. Endorphins generated by exercise can reduce anxiety, tension, anger, and confusion.[16] Although endorphins are often credited with the feeling of calm and well-being people sometimes experience with prolonged exercise, recent research has found that exercise can increase the body's production of endocannabinoids, which may also account for these effects.[17] Cannabinoids, which are small lipids that stimulate cannabinoid receptors to reduce pain and anxiety and generate a sense of well-being, are one of the ingredients that give marijuana its potency. Laboratory animals that have been rendered incapable of making cannabinoid receptors have been observed to exercise less, and they don't appear to enjoy exercise as much.[18]

Epinephrine

As discussed in chapter 3, epinephrine opposes the action of insulin, mobilizes glycogen, and is the chief mobilizer of fat stores.[19] Fat stores (in the form of triglycerides) in the body's fat cells are constantly being broken down and built up. Most of the fatty acids that are released from triglycerides cycle back into fat stores as long as glucose and glycerol in the cells is abundant. When glucose and glycerol are sufficiently depleted, though, more fatty acids are released into the blood to be used as fuel. Thus, the glucose level inside fat cells is a major factor in determining how extensively fatty acids are released into the blood.[20] The sensation of hunger signals a particularly effective time to fastercise because it promotes the increased release and use of fat when the body is already inclined in that direction. The mobilization of stored glycogen and fat with one minute of fastercise can satisfy hunger for as long as two hours (when glycogen stores and hydration are sufficient).

One minute of high-intensity exercise such as fastercise can raise epinephrine levels dramatically (on the order of tenfold) while twenty minutes of jogging may produce little change.[21] Research shows that epinephrine secretion goes up significantly with intensity and power (muscle tension and speed).[22] Epinephrine secretion also increases with duration so that one minute of high-intensity exercise like fastercise will drive up epinephrine levels more than six seconds does.[23] Secretion also increases as the mitochondrial process becomes overwhelmed. When the mitochondrial process is 75 percent maxed out, epinephrine levels begin to increase exponentially and continue to increase once the mitochondrial process is overwhelmed. Epinephrine levels also rise more quickly in people who do high-intensity exercise regularly. (The more we move the survival balance toward forage mode and away from storage mode, the more our bodies adapt to support our capacity to hunt, forage, and escape.)[24]

Epinephrine signals an amplifying cascade of enzymes called the cyclic-AMP cascade. Epinephrine binds to a receptor on muscle cell membranes, which leads to the formation of cyclic AMP (cAMP), a type of messenger molecule. The cAMP activates an enzyme called protein kinase A. This enzyme in turn activates many molecules of phosphorylase kinase, each of which activates many molecules of glycogen phosphorylase, each of which breaks down glycogen molecules to form glucose. Thus, the signal received by the binding of a small number of epinephrine molecules is amplified and generates a fast, large release of glucose.[25]

High-intensity exercise like fastercise increases the release of epinephrine into the blood, and the more intense the exercise, the greater the release.[26] Studies show that high-intensity exercise provides better results with training over time (ten weeks, for example).[27] Thus, it is likely that the more regularly we fastercise, the bigger our epinephrine responses will become, and the better results we'll see.

Thyroid Hormone

Thyroid hormone helps the body respond effectively to cold and danger. Epinephrine and thyroid hormone are synergistic.[28] They work together to help us burn fat, keep warm (maintain normal body temperature), and escape dangerous situations. Thyroid hormone increases the levels of several of the same key mediators involved in the cAMP cascade. By increasing the key mediators—from receptors to some downstream kinases—thyroid hormone prepares cells to respond vigorously to epinephrine. Thyroid hormone also increases the density of epinephrine receptors on fat cells.[29] Thus, adequate thyroid function supports everything we are trying to accomplish with fastercise, including the mobilization and burning of fat and increasing the size, function, and number of mitochondria.[30]

At the same time, epinephrine promotes the action of thyroid hormone. Epinephrine increases the conversion of the less-active form of thyroid hormone (T4) to the active form (T3) by a deiodinase enzyme (especially in muscle, brain, brown adipose, and other tissues).[31] Epinephrine may also increase the number of active thyroid hormone receptors by decreasing their destruction. Since fastercise stimulates epinephrine release, which promotes the action of thyroid hormone, it follows that fastercise can increase the body's metabolic rate and body temperature. This is the science that underlies shivercise—an action the body does automatically in order to raise metabolic rate and body temperature.

When our thyroid function is too low, we can't respond adequately to epinephrine to burn enough fat to stay warm. Our body temperatures run low and we tend to gain fat. Conversely, excessively high body temperatures due to hyperthyroidism or ingestion of certain drugs, such as the amphetamine "ecstasy," are due to the effects of excessive epinephrine and thyroid hormone.[32]

Many of the symptoms of hyperthyroidism are identical to the symptoms of epinephrine overdose, including pounding, fast, or irregular heartbeat as well as nervousness, anxiety, or restlessness. It appears that by increasing many key intermediates, thyroid hormone potentiates the cascades that epinephrine triggers. This explains why very small amounts of exertion, stress, or surprise can generate exaggerated symptoms in people with hyperthyroidism, which includes elevated body temperature. In chapter 10, I discuss more about thyroid function and why it plays such an important role in sustaining healthy body temperature and metabolism, and how fastercise may have beneficial effects on thyroid function.

Growth Hormone

Growth hormone is tremendously helpful in burning fat and building muscle and affects practically all the tissues of the body, as I explained in chapter 3. As one of the counterregulatory hormones to insulin, GH helps increase blood sugar levels when they drop too low. However, when glucose levels go up, GH levels go down because there is no need to free up more glucose. Obese people tend to have higher insulin levels and don't secrete as much GH in response to fasting and exercise. Research suggests that insulin is a major determinant of the reduced GH response in obese people. That's unfortunate because it means that being obese makes it harder to burn fat and build muscle. Thus, high insulin and low GH pave the road to Obesity while lower insulin and higher GH pave the road toward Leanness.

One approach to helping people who are deficient in GH is to dose them with GH artificially. Doctors sometimes try to augment the body's production of GH by giving GH-deficient people injections of growth hormone in an effort to help them lose weight. Doctors may prescribe the shots to be given weekly, several times a week, daily, or even several times daily, and they can be quite expensive. The thing is, our bodies generate GH numerous times each day at different times in response to different signals. GH secretion goes up when we enter the non-fed state and when we exercise, for example. That's handy because GH increases fat breakdown more when it doesn't have to compete with high insulin levels that are trying to store fat. If the body produces too little GH, it can lose muscle and gain fat, developing sarcopenic obesity. Excessive dosing of GH (by injections) can also lead to sarcopenic obesity (via insulin resistance). This is another example of the importance of balance. We don't want too little GH and we don't want too much. Under the right circumstances, the body generates

The Testosterone Effect

Boosting testosterone is an indirect mechanism by which high-intensity exercise like fastercise promotes fat loss and muscle gain. High-intensity exercise increases testosterone,[33] and testosterone increases growth hormone secretion, which increases muscle growth.[34] Actually, both growth hormone and testosterone help build up muscle size and strength, but the effects of growth hormone are probably more important in this regard than those of testosterone.[35] It seems that muscle begets muscle, because the more muscle the body has the more testosterone it makes.[36] Conversely the more fat in the body, the less testosterone is present, because fat cells produce the enzyme aromatase, which converts testosterone into estradiol (an estrogen).[37] When excess fat is lost, aromatase levels can drop and testosterone levels can return to normal.[38]

just the right amount of GH at the right time to maximize fat loss and muscle sparing. However, trying to inject enough, but not too much, growth hormone at optimal moments to synchronize with the body's circumstances would be very hard to do. Fortunately, we can try to stimulate the body's natural production of GH at just the right times in just the right amounts by answering hunger with fastercise. This is a sustainably easy, inexpensive, and convenient way to get perfectly natural dosing of GH.

Growth hormone increases significantly with high-intensity exercise like fastercise but it does not do so with one hour of moderate-intensity exercise.[39] Immediately following high-intensity exercise, the breakdown products of fat and glycogen increase substantially while the breakdown products of muscle markedly decrease.[40] As we head toward leanness and become less obese, our GH response will improve as will our ability to lose fat and gain muscle.

Both GH and lactate help with wound healing and skin repair by increasing collagen deposition and new blood vessel growth.[41] Conveniently, fastercise

also increases lactate. The connection between GH, lactate, and wound healing makes sense considering that escaping dangerous situations and foraging
for food might increase our chances of picking up more nicks and scrapes.

Ghrelin

As discussed in chapter 3, ghrelin brings on the sensation of hunger, and
levels increase when we don't eat and decrease when we do. Ghrelin turns
down energy consumption (probably by decreasing thyroid hormone
expression) and decreases fat oxidation. High-intensity exercise such as
fastercise counters all of that.[42]

Epinephrine can suppress ghrelin, and the higher the level of epinephrine
in response to exercise such as fastercise, the more ghrelin is suppressed.[43]
On the other hand, moderate- to low-intensity exercise does not suppress
ghrelin and may increase it, especially in women.[44] Importantly, while
ghrelin decreases with all forms of intense exercise, it doesn't rise higher
afterward to compensate, not even the following morning.[45] Conversely,
choosing not to eat and not exercise increases ghrelin.[46]

Some studies show that the suppression of ghrelin with high-intensity
exercise lasts for about 30 minutes. However, I believe that the suppression
of ghrelin and hunger can last from 30 to 120 minutes or even longer
depending on how much glycogen we have stored, how intensely we
fastercise, and what type of diet we're on. Nevertheless, the suppressive
effective eventually does wear off, and when hunger comes back, we can
again choose to answer with food, fastercise, or neither.

Ghrelin is one of the primary stimulators of the release of growth
hormone, but high-intensity exercise like fastercise can significantly
increase growth hormone levels independent of ghrelin.[47] Studies have
shown that moderate- to high-intensity exercise increases GH without
increasing ghrelin, which indicates that ghrelin was not responsible for the
GH increase.[48] Thus, when the energy from a meal runs out, ghrelin levels
rise, causing appetite and GH to increase, but with high-intensity exercise,
the exercise itself triggers GH to increase, which in turn causes ghrelin and
appetite to go down. It's as though high-intensity exercise is the antidote
for the effects of going hungry.

One study showed that while continuous cycling and treadmill usage
for thirty to sixty minutes didn't lower ghrelin, shorter ten-minute bouts

of increasing intensity did suppress ghrelin release, and the degree of suppression was proportional to the intensity.[49] This also relates to the phenomenon by which fastercise makes hunger quickly go away. In this study, ghrelin suppression coincided with increased epinephrine and norepinephrine. (Norepinephrine is nearly identical in structure to epinephrine.) Epinephrine and norepinephrine decrease blood supply to the stomach, which is a source of ghrelin. When a fatty acid (acyl group) is attached to ghrelin, the result is acylated ghrelin, which is the only form responsible for releasing GH and increasing appetite. Acylated ghrelin suppression increases according to the intensity of exercise and is suppressed more in the presence of growth hormone.[50] This may help explain why some people are satisfied with smaller meals immediately after intense exercise. Acylated ghrelin is suppressed even more when we exercise in a warm environment, presumably because the body doesn't need as much fuel to stay warm.[51]

If high-intensity exercise *lowers* ghrelin and appetite, where does the expression "working up an appetite" come from? It may express the response of the body to low-intensity exercise, which does result in an increase in ghrelin (and appetite).[52] Many people will go to the gym and run on a treadmill or ride an exercise bike for forty-five minutes or more and will become very hungry soon thereafter. This holds for outdoor activities as well, particularly in cold environments (chopping wood, ice-skating, cross-country skiing). It's easy to work up an appetite performing such activities, because when done at low-intensity, they cause an increase in ghrelin. And remember, cold temperatures may bolster that effect, just as a warm environment intensifies ghrelin suppression.

In addition to low-intensity exercise, other factors can also increase ghrelin levels. Sleep deprivation can lead to higher ghrelin levels. Stress can increase ghrelin levels, which might explain why some people tend to eat when under stress as well as why they tend to crave energy-dense or "comfort" food when under stress. This effect of stress can be countered with fastercise.

Cortisol

High-intensity exercise like fastercise increases levels of epinephrine, growth hormone, and cortisol, but one hour of low-intensity exercise does not.[53]

Cortisol levels also increase with exercise intensity and can remain elevated for two hours.[54] This is beneficial because we need cortisol to permit the full effect of glucagon and epinephrine activity.[55] Following high-intensity exercise, cortisol declines to an even lower level than prior to exercise.[56] This is handy because, as noted in chapter 3, persistently elevated cortisol levels can lead to fat gain and other health problems. For example, prolonged cortisol elevation and stress can contribute to persistent (rather than normal transient) elevations in blood glucose, which can be especially problematic for diabetics.[57] And prolonged cortisol elevation can lower growth hormone levels and contribute to muscle wasting (as in Cushing's disease). That is, prolonged stress can hamper fat loss and muscle growth.[58]

Since cortisol level increases for an hour or two after high-intensity exercise and then declines afterward, I believe that fastercise can help encourage the variability that is desirable in cortisol levels. Notably, short bouts of intense exercise (like fastercise) have not been shown to reduce immune function, whereas ninety minutes of intense exercise may lead to suppressed immune function from several hours to several days later, and even longer following the running of a marathon.[59]

Insulin

I discussed insulin at length in chapter 3, so here I'll focus on specifics relating to the effects of exercise on this all-important hormone. Lowering glycogen stores is a good way to promote insulin sensitivity. Both fasting and high-intensity exercise like fastercise reduce glycogen stores and improve insulin sensitivity. High-intensity exercise can reduce insulin resistance (23–58 percent) and is great for fat loss and aerobic (mitochondrial process) and anaerobic (cytosolic process) fitness.[60] One study showed that a program of fifteen weeks of high-intensity exercise three times per week significantly reduced total body fat, subcutaneous leg and trunk fat, and insulin resistance in young women.[61] High-intensity exercise such as fastercise can also improve pancreatic insulin production in people with type 2 diabetes.[62] Even a single exercise session can improve insulin sensitivity in people who are healthy or prediabetic or who have type 2 diabetes.[63] Thus, as insulin sensitivity improves and excessive insulin diminishes, fastercise can continue over time to shift the body's survival balance away from storing fat and more toward building muscle.

Energy Pathways for Exercise

At the heart of obtaining the benefits of answering hunger with fastercise are the energy pathways for exercise. These pathways generate ATP, which is the energy currency of the body. I introduced these pathways in chapter 2: the cytosolic and mitochondrial processes of cellular respiration. The mitochondrial process encompasses the citric acid cycle (CAC) and oxidative phosphorylation (OXPHOS). Exercise calls on the body for an increased supply of energy to sustain the physical effort. Here I examine these pathways that generate ATP and the related metabolic reactions that take place during exercise and while the body recovers after an exercise session.

Fastercise and ATP Production

Fastercise involves intense muscle activity, and ATP provides the energy for that activity. Every time muscles use a molecule of ATP, two protons are generated. The production of CO_2 during the CAC also generates protons. The more protons in a solution, the more acidic it is. Thus, when ATP releases energy to power our muscles, they become more acidic.

As described in chapter 2, the cytosolic process quickly produces ATP as well as pyruvate, which feeds into the mitochondrial process to produce even more ATP. The cytosolic process also predigests and energizes glucose into lactate for use by our brain, heart, muscle, and other tissues. During strenuous exercise, the mitochondrial process can fall behind as if falling into debt. When the mitochondrial process cannot keep up with the pyruvate generated by the cytosolic process, the excess pyruvate remains in the cytosol and receives protons from NADH and acid to form lactate. Thus, as fastercise overwhelms the mitochondrial process, lactate can build up quite quickly. At the same time, acid levels in the muscles are increasing due to the utilization of ATP. Because lactate and acid are building up simultaneously, it's common to refer to this as a buildup of lactic acid. However, lactate is not an acid, it's a base.

During highly vigorous exercise such as fastercise, acid can build up so quickly that we begin to experience a burning sensation in our muscles, and our muscles simply can't keep up the intense work for more than a minute or two. At such a time, respiration rate increases to expel carbon dioxide, which has the effect of clearing acid from the muscles. (Carbonic acid combines with a proton to form CO_2 and H_2O, so getting rid of CO_2 is a way of getting rid of acid.) We start breathing faster and then deeper and start feeling winded. When we fastercise intensely enough, the mitochondrial process becomes overwhelmed and even deep breathing (gasping) isn't enough to prevent acid from building in the muscles. The acid can build up so high that we must slow down or stop, whether we want to or not.

The usual ratio of ATP to ADP (adenosine diphosphate) in the body is about 500:1, but intense exercise can quickly lower this ratio, sending a powerful signal that the body needs to dramatically increase ATP production. In order to quickly replenish ATP, our bodies use phosphocreatine (PCr) as a reserve of high-energy phosphates to convert ADP back into ATP. (This is a different process than OXPHOS.) Our bodies normally keep more PCr on hand than they do ATP. The body breaks down PCr and transfers a high-energy phosphate group to ADP (which has two phosphate groups) to form ATP (which has three phosphate groups). The breakdown of PCr increases right when exercise begins and peaks within eight to twelve seconds, depending on the intensity of exercise and other factors. This has fascinating real-world implications. Currently, the fastest runners in the world can run one hundred meters from a stationary start in less than ten seconds. These runners typically reach their top speeds and then begin to slow down about eight seconds after the start of the race. The eight-second point would be about when PCr utilization would peak, and the runners' bodies would need to start depending more heavily on other, slower pathways for ATP production. Thus, it's possible the slowing of their speed may be linked to this shift in energy pathways.

During the first ten seconds of all-out exercise, about half of the energy used comes from ATP that was formed through phosphorylation by PCr and about half comes from the cytosolic process. After a race (simulated by a bout of fastercise), it typically takes about fifteen minutes to replenish PCr stores, depending on O_2 supply and the capacity of the body's acid-clearing systems (mitochondria, breathing off CO_2, kidneys).[64]

Clearing Acid and Lactate

Acid buildup causes muscle fatigue, and so does lactate, even without a drop in pH.[65] Because of this, after exercise we need to clear the buildup of acid and lactate in order renew our strength. When we stop fastercising, the mitochondrial process catches up and lactate and acid levels decline. The mitochondrial process helps clear acid by passing high-energy electrons to oxygen and protons to form water. Many of those electrons come from NADH generated by the malate shuttle, which is a pathway that increases the amount of NADH within the mitochondria while increasing the amount of NAD (NADH minus a proton) in the cytosol. That's super handy because the NAD scavenges back protons, converting lactate to pyruvate. The pyruvate can then enter the mitochondrial process. Lactate also can be used as fuel by other tissues of the body and can be transported to the liver and converted back into glucose. The cycling of glucose into lactate in the muscles and some of that lactate back into glucose in the liver is known as the Cori cycle. This recycling of lactate back into glucose by the liver saves the muscles from having to clear all the lactate on their own. In the Cori cycle, the cytosolic process generates two ATP from each molecule of glucose processed to lactate. Some of that lactate is transported through the blood to the liver where six ATP are used to convert lactate back into glucose, which can be recirculated into the bloodstream. That's a net loss of four ATP per glucose molecule per turn of the Cori cycle. The body is willing to expend this ATP in order to clear lactate from the muscles to help preserve their strength and also to help replenish blood glucose to fuel brain function. Thus, fastercise burns a lot of energy in a short period of time generating lactate, and the processing of that lactate back into glucose also burns a lot of energy.

The buildup of lactate and acid and the depletion of ATP and PCr is sometimes referred to as an oxygen debt. Oxygen debt is thought of as the amount of oxygen that needs to be processed by the mitochondria in order to return lactate, acid, ATP, and PCr to their preexertion levels.

The current understanding of the role of lactate has changed dramatically from past views. Lactate is produced under aerobic (rest and low-intensity exercise) and anaerobic (high-intensity exercise) conditions. It is a signaling molecule and a valuable fuel we use while at rest as well as during exercise. Even at rest, half of the lactate produced in the cell is burned as fuel, and half is sent to the liver to be recycled into glucose. Our bodies can replenish muscle glycogen using lactate both directly and through the Cori cycle, even in the absence of food.[66]

Why the Herd Yawned

I have noticed that when I answer hunger with tightercise and my hunger goes away, about ninety seconds later I have a tendency to yawn. I picture my contracting muscles using up ATP, releasing stored fuel, increasing my CAC intermediates, encouraging the CAC to turn, and OXPHOS gearing up to process high-energy electrons to replenish ATP. Because those high-energy electrons are passed to oxygen, which then forms water molecules, and because my cells are generating CO_2 and clearing acid, it makes sense to me that about ninety seconds later, my body responds by yawning to blow off that CO_2 and take in more oxygen.

This may also help to explain why yawning is an adaptive behavior that spreads from one individual to another (why yawning is "contagious"). In a herd of animals, one animal might spot a hunting opportunity or a danger that the rest of the herd doesn't see. The alerted animal instinctively stretches and yawns, which gears up the CAC and OXPHOS to generate energy, preparing its body for imminent physical activity. The rest of the animals instinctively begin stretching and yawning also, because if one herd animal needs to prepare for physical activity, it is likely the rest will need to leap into action, too. We can add this to the many theories on the purposes of yawning.

During exercise, even more of the lactate is burned as fuel (75–80 percent).[67] This field of research is emerging and our understanding will no doubt deepen over time. But for now, we know that there are transporters (monocarboxylate transporters) on cell membranes that help clear built-up lactate and acid by transporting lactate and protons out of cells. There are different types of monocarboxylate transporters. Monocarboxylate transporter 1 (MCT1) not only exports lactate but also imports lactate into cells and mitochondria, where the lactate is converted by lactate dehydrogenase (LDH) to pyruvate, which is burned to make ATP.[68] Because this mitochondrial process uses up protons, it also clears acid. Although most of the lactate is cleared by the exercising muscle fibers themselves, excess lactate is exported from exercising muscle fibers and imported by nonexercising muscle fibers both nearby and distant. Regular fastercise builds up the size and number of mitochondria, the concentrations of MCT1 (on the muscle cell membrane as well as the mitochondrial membrane), Na+/H+ exchanger proteins, and a portion of the OXPHOS pathway, all of which help process oxygen and clear lactate and acid faster.[69]

Training increases lactate clearance both locally and systemically, giving the body quicker and easier access to more energy. Better processing of lactate gives the body more access to the cytosolic process (because lactate inhibits glycolysis) and better processing of oxygen gives the body more access to the mitochondrial process.[70] This enables us to exercise longer with more energy and less fatigue and without getting as winded. Regular fastercise can improve our blood circulation and help us hold our breath longer. It can even help people with lung disease to breathe easier.[71]

Let's consider the role of the Cori cycle in intense exercise when only about 20–25 percent of the lactate generated by the muscles is sent to the liver for recycling to glucose. Even though that's a small percentage of lactate being converted back to glucose, this process consumes a substantial amount of energy, because there is a net loss of four ATP for every glucose molecule that completes the Cori cycle. That's like borrowing $200 from the bank one minute and the next minute paying it back in full with an additional $400 in interest! This process just goes to show how important it is for the body to be able to escape dangerous situations and for the brain to have the glucose it needs. The body is willing to pay a huge interest rate in order to have access to the energy it needs immediately. This process may also shed light on the phenomenon that people lose more fat by engaging in high-intensity exercise than they do in steady-state exercise.[72] When we fastercise, fatty acids are released from our fat stores for energy, and lactate is released from our muscles. It appears that in some ways lactate can inhibit the burning of fat and in some ways it can promote it.[73] Either way, lactate is cleared fairly quickly after fastercise because the exercise itself is such a brief bout of activity.

To summarize: When we fastercise our bodies first use up the ATP and PCr on hand. The ATP is depleted in two to three seconds and the use of PCr to replenish ATP peaks in about eight to twelve seconds. After that, our cells are obligated to rely more heavily upon the cytosolic and mitochondrial processes to generate more ATP. The mitochondrial process is quickly overwhelmed and can't process pyruvate as fast as the accelerated cytosolic process is producing it. ATP use generates acid faster than the mitochondria can clear it, and acid builds up. Blowing off CO_2 increases dramatically and we start to become winded within fifteen to twenty seconds. Nearly a minute into fastercise we can notice our body heating up (close to breaking a sweat). We find our muscles may start to burn and we can catch a deep breath. At this point, we have to slow down or stop.

Fastercise isn't so much about how many calories we burn in the moment as it is about signaling the body to lower its fat set point, burn stored fat instead of food, and speed the metabolism. We fastercise intensely until we can catch a deep breath so that we can overwhelm our mitochondrial process in order to stimulate mitochondrial biogenesis and increase excess postexercise oxygen consumption (increase our metabolic rate). Although fastercise can mobilize glycogen to push off hunger, the more we fastercise, the quicker we use up our glycogen, ketones, and CAC intermediates, which can lead to our getting hungry again sooner. This is probably why our morning stretch when we wake up only lasts for several seconds. That's enough time to stimulate the breakdown of glycogen and fat to fuel our CAC, but not enough time to deplete the supply of phosphocreatine or to deplete much fuel in the mitochondria.

Training the Energy Pathways

Stepping up from the effects of fastercise on energy pathways at the cellular level, let's consider the effect of fastercise on muscle tissue and adipose tissue (body fat). The body has two types of muscle fibers: fast-twitch fibers (also called white muscle fibers) and slow-twitch fibers (also called red muscle fibers). And there are two types of adipose tissue: brown adipose tissue and white adipose tissue.

High-intensity exercise such as fastercise involves the fast-twitch muscles, which consume energy faster than the slow-twitch muscles used in endurance activities.[74] Fast-twitch muscles and the browning of adipose tissue (which I discuss later in this chapter) help account for why high-intensity exercise often generates much more fat loss than lower-intensity exercise.[75]

Building Muscles

When we approach the limits of athletic performance, it appears that different training and adaptations are optimal for different events. For example, a top sprinter is no match for a top marathoner in the marathon. But a top marathoner is no match for a top sprinter in a sprint. Also, a top middle-distance runner can easily defeat a top sprinter and a top marathoner in a middle-distance race. The adaptations involved are both molecular and structural. For example, sprints call for large amounts of energy in a short time period. Sprint activities place heavy demands on both the cytosolic and mitochondrial processes. Fast-twitch (white) muscle fibers are well adapted for that, because they produce plenty of the enzymes involved in the cytosolic process, which provides power when the mitochondrial process is overwhelmed.[76]

On the other hand, to support the consumption of oxygen by the mitochondrial process, slow-twitch (red) muscle contains a lot of myoglobin (a type of protein molecule that carries oxygen), mitochondria, and capillaries. Slow-twitch muscles are "blood red" because both myoglobin and hemoglobin (a blood protein) have a heme group, which contains iron. Endurance activities rely heavily on the mitochondrial process and not so much the cytosolic process. Red muscle fibers are excellent for providing maximal power for longer periods of time in endurance events. This helps explain why the breasts of chickens are composed of white muscle fibers (white meat). The contractions of white muscle fibers are faster and deliver greater force, but they are harder to repeat with as much endurance as those of red muscle fibers. White muscle fibers can supply the rapid bursts of energy chickens need for short flights. Chicken thighs and legs are composed of dark meat (more red muscle fibers) to provide more endurance for running. On the other hand, duck breasts are composed of dark meat to provide endurance for long migratory flights. White muscle tends to increase in size with training. Thus, sprinters tend to have bigger muscles than middle-distance runners and marathoners.

We can choose the training we want to do based on the adaptations we want. For example, because shivercise helps build up both the cytosolic and mitochondrial processes, shivercise can enhance performance in both sprinting and endurance events. Lactate can be generated by white muscle fibers and transported and cleared by red muscle fibers, so high-intensity activities such as fastercise encourage the growth of both white and red muscle.

To build white muscle fibers as well as red muscle fibers we need to overwhelm our mitochondrial process by demanding more speed or power than either can deliver. We can do that in less than a minute of fastercise (especially tightercise). And our bodies can become bigger, stronger, and faster with regular fastercise.

Improving Endurance

Our bodies combine fuel with oxygen in the mitochondria to generate energy, but that oxygen first has to be transported from the lungs through the blood and ultimately into the mitochondria. Thus, pulmonary, cardiovascular, and mitochondrial health and function are all important for endurance.

Fastercise improves both cardiovascular and mitochondrial health and function. High-intensity exercise like fastercise results in increased levels of AMP-activated protein kinase (AMPK), an enzyme that can stimulate the refurbishing of existing blood cells as well as the formation of new blood cells.[77] In addition, AMPK stimulates an activator molecule that stimulates an increase in the size and number of mitochondria, promotes the number and function of endurance muscle fibers, and increases fat burning.[78] It appears that this activator molecule is stimulated by high-intensity exercise through one biochemical pathway and by lower-intensity exercise through another biochemical pathway,[79] which explains how it's possible to get similar benefits from a minute or two of fastercise as compared to much more time on a treadmill. In *The One-Minute Workout*, prominent exercise researcher Martin Gibala describes how Czechoslovak athlete Emil Zátopek relied on high-intensity training almost entirely to prepare for the 1952 Olympics. Zátopek won gold medals not only in the 5,000-meter and 10,000-meter events but also in the marathon event, even though he had *never run a marathon before*. Of course, Zátopek did train for more than just a few minutes a day, but still!

Although high-intensity and continuous exercise both stimulate some similar adaptations, they also stimulate different ones. Therefore, one recommendation for modern elite endurance athletes is to do a combination of about 75 percent low-intensity training and about 10–15 percent very-high-intensity training.[80]

Converting Fat to Heat

One reason why exercise is one of the main interventions for the treatment of obesity is that it increases the energy expenditure of a type of fatty tissue called brown adipose tissue (BAT). Brown adipose tissue (BAT) is a distinctly different type of adipose tissue than white adipose tissue (WAT). WAT functions to *store* fat for use as fuel to make ATP; BAT functions to *use* fat to generate heat to keep us warm. BAT is brown in color because it contains far more mitochondria (which contain iron), as well as a higher concentration of tiny blood vessels. It was once believed that BAT was present only in newborn babies. Recently, however, BAT has been found in human adults, which has led to research about avenues to manipulate BAT to combat obesity and metabolic disease. It's easy to see how increasing BAT's conversion of fat to heat would be helpful in the treatment of these chronic conditions.

The number and function of BAT cells in the body can change under various conditions. Also, even though BAT cells and WAT cells remain distinctly different, under certain conditions, WAT cells can become more like BAT cells by developing many more mitochondria and by functioning more like BAT cells. When they do, it is said that the WAT cells have become "beige" or "brite." When conditions revert to normal, beige cells can become white again. Similarly, when conditions call for less BAT function, BAT cells can function more like WAT cells.[81] For example, when we are exposed to cold temperatures, both our BAT and WAT cells can develop more mitochondria and become more brown. At the same time, the number of BAT cells can increase as well as the quantity of enzymes used in fat transport and fat burning in the mitochondrial process.[82] All of this leads to increased fat burning and heat production.

A 2013 study presented in the journal *Cell Metabolism* has shown that exercise can generate similar changes in adipose tissues.[83] BAT function is often called nonshivering thermogenesis, because it refers to heat generated by fat cells not by muscle cells. But that doesn't mean that shivering and nonshivering thermogenesis are unrelated. In fact, it appears that they are very much related. When we get cold we tend to shiver, and the more intensely we shiver, the more our muscles produce molecules that signal the increased browning of adipose tissue. Exercise causes muscles to produce those signaling molecules just as effectively as shivering does.[84] This suggests that it's the muscle activity of shivering that leads to the browning of fat, whether that shivering is involuntary due to the cold or voluntary due to fastercise. This makes sense, because fastercise and other high-intensity exercise increase the size and number of mitochondria in the muscles.[85] Research also indicates that high-intensity exercise like fastercise can increase BAT function and the number of mitochondria in fat cells.[86] Numerous studies have demonstrated that people with higher amounts of BAT have better metabolic fitness (less obesity, insulin resistance, diabetes).[87] One way that fastercise helps to shift the survival balance away from fat storing and more toward fat burning is through the browning of adipose tissue. The more fastercise the more BAT; the more BAT the more fat burning and metabolic fitness.

The Physiology of Shivercise

Shivercise is fascinating because of the ingenious mechanism of muscle contraction. At the microscopic level, muscle contraction is dependent on cross-bridges that form between contractile filaments in muscle cells. These filaments are composed of proteins called actin and myosin. Myosin changes shape when it "grabs" actin, creating a cross-bridge that generates mechanical force. After the myosin grabs the actin, it very quickly releases it. Energy from ATP is used to reset the myosin to its original shape so that it can form another cross-bridge. Thus, the formation of every cross-bridge generates mechanical force (like a minicontraction) that lasts a very short time, and all the minicontractions together manifest as a full-scale muscle contraction. The entire process of forming a cross-bridge, minicontracting, releasing, and resetting to grab again can take less than two-tenths of a second, even faster in fast-twitch white muscle.

The amount of ATP (energy) needed for skeletal muscle contraction is dependent on the force, duration, shortening, and velocity of the contraction and the length of the muscle.[88] Our muscles generate more force by generating more cross-bridges. Let's consider this through a simple everyday example of muscle contraction. When you pick up a glass of water and hold it in the air, a certain number of cross-bridges form in your muscles. Since cross-bridges expire over time, your muscles need to keep forming new cross-bridges in order to keep holding up the glass. This involves using ATP to reset myosin, so even though the glass itself isn't moving, your muscles are consuming energy. Picking up a ten-pound dumbbell and holding it in the air would require even more cross-bridges and more ATP because more force is needed to hold up the dumbbell than a glass of water.

Even *more* cross-bridge recycling is needed to cause a muscle to shorten, rather than simply maintaining a stationary position. And the faster a muscle shortens, the more cross-bridges are needed. A rapidly shortening muscle can use *three times* more ATP than one that is not shortening! Thus, raising and lowering a dumbbell requires more ATP than simply holding a dumbbell still.

The goal with shivercise is to generate and break as many cross-bridges in our muscles as fast as we can in order to burn ATP as quickly as possible to overwhelm the mitochondrial and cytosolic processes. By overwhelming the mitochondrial and cytosolic processes, shivercise stimulates the body to increase the size and number of mitochondria. I believe that one of the aspects of shivercise that leads to heavy consumption of ATP is the rapid changes of direction of the body parts in motion. Runners use up a tremendous amount of energy at the very start of a sprinting race because they are starting from a dead stop. Airplanes use more energy taking off and climbing to cruising altitude because they have to fight gravity. Sure, the faster we run, the more energy we burn. But imagine how much more tiring it would be to run as fast as you can while having to fight inertia the whole time by sprinting a few yards, coming to a full stop, turning around and doing the same thing in the opposite direction? And doing that over and over again. Start. Stop. Start. Stop. In sports training, that kind of exercise is so grueling it's called running "suicides." How would that look on the level of an individual muscle? It would look like shivering.

The more muscles we involve—especially the long muscles such as our thigh muscles, biceps, triceps, and back muscles—the better. The more muscles that repeatedly contract and release as fast as possible, the more quickly our bodies burn ATP and generate heat. Have you ever noticed that shuddering and shivering seem to involve many of your muscles? That's to mobilize as much energy as possible in order to stay warm. And the greater the length of the muscle involved, the more cross-bridges are needed.

Shivercising uses up ATP so quickly that a person can completely overwhelm the mitochondrial and cytosolic processes in their muscles in about one minute. The faster you move, the more profoundly you'll cancel your hunger (and perhaps longer), preserve muscle, burn fat, generate heat, and build fat-burning machinery. Have you ever been so cold that you started to shiver violently? If you have, you probably remember it as a pretty serious moment. It's an alarming situation for the body to become so cold. There are few things more critical for your survival than a healthy body temperature. And when yours gets too low, what does your body do? It shivers, uncontrollably if need be, and you start breathing more deeply. It will do what it needs to do to speed your metabolism and raise your body temperature. The rapid movement during shivering is your body's most efficient mechanism for warming itself to maintain a healthy body temperature. I believe that shivering is the activity that also most strongly signals the body to increase the size and number of its mitochondria to prepare it to more easily handle exposure to the cold in the future. Fortunately, we can voluntarily shiver to send the body that signal without having to suffer the experience of getting seriously chilled. This is great news, especially for people who feel as though they are freezing all the time, even at normal room temperature. Regular shivercise may reduce this tendency over time.

High-intensity exercise such as shivercise increases our metabolic rate for a time even after we stop the vigorous movement. This effect is called excess postexercise oxygen consumption (EPOC).[89] Although shivercising may inhibit fat burning for a minute or two to promote the clearing of lactate, it is likely that the resulting release of fatty acids and increase in EPOC fat burning more than makes up for it.

The Physiology of Tightercise

Tightercise is a type of isometric, or fixed-length, contraction of muscles. The amount of ATP (energy) used up during tightercise depends on the length and size of the muscles involved, the force of the contractions, and the duration. Muscle cells act as a unit, and they usually contract at the same time as other muscle cells that are activated by the same motor nerve. A group of muscle cells and its motor nerve is known as a motor unit. As we've discussed, interactions between actin and myosin generate the force of contraction in muscles. These interactions last less than two-tenths of a second. At any given moment, some cross-bridges are letting go while other cross-bridges are forming. Thus, our muscle contractions are made up of innumerable overlapping contractions that enable us to hold a contraction steady, at least at first. As we continue to hold an intense contraction, some of our motor units become fatigued and give out, leaving the force to be maintained by a smaller number of motor units. With fewer and fewer motor units to overlap and smooth out the load, the contraction becomes more unsteady, and our muscles start to shake. Fixed-length muscle contractions can generate greater force than contractions that shorten muscles. Greater force stimulates more muscular growth and strengthening of the tendons.[90] The more tightly you contract your muscles, the stronger the signal you send your body. As you contract your muscles with progressively greater force day by day and week by week, they are stimulated to grow. When you ball up your muscle in a full contraction, the actin and myosin filaments in your muscle overlap more, providing an opportunity for more cross-bridges to form. This is why tightercising your muscle in the balled-up position may be most effective especially for building size and strength.

Tightercise is an extremely efficient form of exercise because it provides a greater percentage of time at peak exertion and many muscles can be contracted at the same time. Tightening your muscles without moving them is quite different from doing repetitive movement exercises. When you do exercises that involve movement, your muscles often experience different amounts of exertion throughout the range of the movement due to the changing angles of your bones and joints in relation to angle of the load. You may reach peak exertion during only a small percentage of the total time you are exercising. Indeed, when working with weights or machines, you may never reach peak exertion at all, because you may be using a load that is too light. Or you may not reach peak effort until the last rep of your last set. With tightercise, however, you can experience peak exertion for almost the entire time you are exercising. You can go straight to peak exertion for a few seconds without even getting up from your desk. Done.

How Can So Little Do So Much?

One minute of fastercise can overwhelm all four of our major energy sources (ATP stores, PCr stores, mitochondrial process, and cytosolic process), but jogging for forty-five minutes doesn't. Mild or moderate exercise doesn't significantly increase epinephrine secretion, but fastercise does.[91] One minute of fastercise releases the counterregulatory hormones as well as chemical-signaling proteins called myokines, which stimulate muscle, liver, fat, brain, bone, and immune cells and may be responsible for many of the benefits of exercise. For example, one myokine called IL-6 stimulates increased glucose production by the liver and increased breakdown of fat while increasing muscle uptake and utilization of glucose.[92] Interestingly, the fast-twitch white muscle cells engaged more in fastercise tend to secrete a different set of myokines than those secreted by the slow-twitch red muscle cells engaged more during endurance exercise. Thus, when we fastercise we aren't just generating mechanical force and movement. In a sense, we are also giving ourselves hormone and myokine injections. Did you know that a small injection of epinephrine can often save a person from dying from a severe allergic reaction? (I wonder if fastercise could help save someone's life if they were having an allergic reaction but didn't have access to a shot of epinephrine.) And a small injection of insulin can save a person from suffering the effects of diabetes. Similarly, small "injections" of hormones and myokines generated from a minute of fastercise can help save us from obesity and muscle loss. By canceling hunger with fastercise we can give ourselves a very appropriate dosage of a mixture of very expensive hormones and myokines at just the right time, for free. Since we can do it for free, we can easily afford to fastercise once, twice, four times, or eight times a day. I think of it as "muscle as pharmacy." Only, it's completely natural and without side effects. Talk about regular exercise—instead of exercising three times a week the way many people do, we can fastercise three to six times a day in less time and with more convenience, less difficulty, and, very likely, better results.

CHAPTER 8

Food Basics

There's no lack of opinion on the "best" way to eat these days. Every month or so, a new book or website or video crops up to explain the newest way to eat right. But as I've explained in earlier chapters, I am not a proponent of any one diet or eating plan. Human beings are very adaptable; we are capable of thriving on a wide variety of diets. And although the dietary needs of all humans are broadly similar, every individual is different, as are our circumstances. Eating habits considered normal in one part of the world can be dramatically different from those considered normal in another.

The industrialization of agriculture has also had a profound impact on modern eating habits. Many critics of the modern food system will tell you that what you eat is determined more by what will boost corporate profit margins than by what will best meet your nutritional needs. And countless diet books full of recommendations on how to change your eating habits aren't necessarily based on sound scientific principles either. Plus, hunger and cravings are always an obstacle to adhering to a prescribed diet plan. But as you've learned from this book, fastercise is a highly effective tool for managing appetite, and canceling hunger with fastercise can help you adhere to any sensible eating plan. If you overeat on a daily basis and choose nonnutritious foods, will fastercise magically make you fit and healthy? No. It's undeniable that what you eat has a huge bearing on your ability to lose fat. But I believe that the combination of fastercise and a healthy diet has the potential to help everyone lose fat, build muscle, and feel better.

Carbohydrate, protein, and fat are the three macronutrients our bodies use for fuel. Macronutrients are types of food that are normally eaten in large quantities in our diets. On the other hand, micronutrients are substances that are needed in small quantities. A healthy diet must include protein, which provides essential amino acids that the human body can't synthesize on its own. Dietary fats are critical to provide essential fatty acids that the body cannot make on its own. Vitamins and minerals are also essential in our diets because they are necessary for our bodies to function properly and our bodies can't make them on their own. We can procure vitamins and minerals from food, and we may choose to take vitamin supplements. Carbohydrates are not essential in the diet in the way that protein and fat are, because the body has the ability to transform protein into carbohydrates. So technically speaking, we can survive without eating carbohydrates, but carbohydrates do play important roles (more about that later in this chapter).

One way to compare diets is by the relative percentages of macro-nutrients they contain. Another way is to compare the source of those macronutrients (plants versus animals, wild versus farmed, refined versus unrefined). How adaptable is the human body to different types of diets? Here's one example: Some people in Africa traditionally eat a diet that is more than 80 percent carbohydrate, while some people who live near the Arctic Circle eat a diet that is about 50 percent fat. The people of Japan have one of the highest life expectancies in the world (higher than that of people in the United States) and they obtain about 60 percent of their calories from carbohydrates.[1] A study involving nearly one hundred thousand participants that investigated the correlation between diet and health in Japanese people indicated that the lives of Japanese people tend to be not only longer than in many countries but also healthier.[2]

It's tempting for proponents of one dietary plan to condemn another as wrong, but different approaches may benefit the same or different people under the same or different circumstances. For example, a recent study involving more than six hundred overweight adults compared the twelve-month weight loss results of a low-fat diet versus a low-carbohydrate diet.[3] The low-fat diet contained 48 percent carbs, 29 percent fat, and 21 percent protein, while the low-carb diet contained 30 percent carbs, 45 percent fat, and 23 percent protein. The subjects were randomly assigned to one diet or the other, and the average weight loss for *both* diets was about thirteen pounds. Some subjects lost more than sixty pounds while a few gained fifteen or twenty. Wow, that's quite a variety of results, and it demonstrates how differently people can respond to a particular diet.

Significantly, the researchers did tell all the participants to buy their food at farmers markets and to avoid eating refined and processed foods. The results of this study suggest that helping people focus on eating healthy and nutritious food may be more of a factor for fat loss than whether the diet is low-carb or low-fat.

We have all heard stories of someone who tried without success a fat-loss diet that had worked beautifully for a friend. The difference might have been due to a hidden factor in their metabolism that the diet recommendations didn't address. It makes sense to try different lifestyle approaches and pay close attention to how your body responds to find out what works best for you, especially on an ongoing basis. And as you discover what's effective, write it down! It's easy to forget details of what you ate and didn't eat last week or last month.

How *Not* to Eat Right

Let's start by considering how we could achieve the goal of *gaining fat* as quickly as possible. Let's say we could take in food only through the gastrointestinal (GI) tract (no intravenous feeding). Remember, we build fat from excess calories, so we would need to ingest lots and lots of calories. Since the GI tract can process only a limited quantity of material at a time, we'd want to eat calorically dense foods that are easy to process and as fattening as possible. We'd want to choose foods that don't generate satiety for long periods. (That way, we'd find it easier to keep eating.) We'd also want to avoid exercise and limit our caloric expenditures. Would you be surprised if I told you that the diet and lifestyle of many people in the United States fits this scenario quite closely? This is part of what has led the United States to one of the highest obesity rates in the world. Over one-third of Americans are obese.

The foods that promote obesity fit the parameters I just described: calorically dense, easy for the body to process, and unlikely to keep you satiated for long. They include sugar, refined carbohydrates, and refined fats.

(Added) Sugar

Sugar amply meets the criteria for an ideal fattening agent: It's calorically dense and provides little satiety. The body processes sugar quickly, and it is uniquely fattening. Plus, sugar is super cheap, so we can afford to buy lots of it. Yikes.

Table sugar is sucrose. A molecule of sucrose consists of a molecule of glucose attached to a molecule of fructose. Sucrose is uniquely fattening because it's fattening in two ways:[4] As described in previous chapters, glucose is fattening because it fills up glycogen stores and stimulates insulin release to promote fat building. But that's only half the story. The other downside of sucrose is that while glucose can be metabolized by all types of cells, only liver cells can process fructose. The breakdown of fructose, as compared to glucose, can lead to higher levels of triglycerides and lower levels of ATP, which can lead to fat deposition and inflammation in the liver.[5] This situation isn't good, because the liver isn't designed to store much fat. Excessive intake of fructose is also known to promote insulin resistance and fatty liver disease. (An abnormal accumulation of fat cells in the liver that can lead to inflammation and even permanent damage.)[6] Thus, eating too much added sugar promotes the insulin resistance that contributes to obesity and also contributes to fatty liver disease. Plus, inflammation of the liver due to fatty liver disease can contribute to insulin resistance, obesity, and type 2 diabetes.

Thus, even though fructose does not raise blood sugar levels much directly, many experts believe that ingestion of fructose contributes more to the progression of type 2 diabetes in the long run than glucose does (even though many people with type 2 diabetes are told that fructose is a safe alternative for them). The main dietary source of excessive amounts of fructose is foods that contain added sugar and high-fructose corn syrup. Fruit naturally contains small quantities of fructose, but it is hard to eat enough fresh fruit to generate the excessive intake that has been shown to promote insulin resistance.[7] However, consuming excessive amounts of fruit, especially in the form of dried fruits and fruit leathers, should be avoided.

The biggest source of added sugar in the diets of people in the United States (accounting for 34.4 percent) is sweetened drinks such as soft drinks, energy drinks, and sports drinks.[8] Other significant sources include grain desserts (12.7 percent), fruit drinks (8 percent), candy (6.7 percent), and dairy desserts (5.6 percent). Grain desserts include cookies, cupcakes, pastries, pies, and I would include some breakfast cereals, too. Commercial

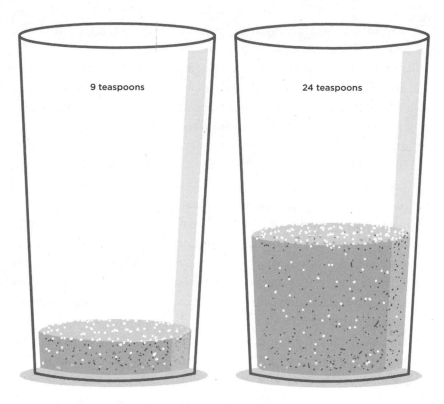

Figure 8.1. A typical twelve-ounce can of soft drink contains nine teaspoons of sugar. Some energy drinks contain twenty-four! This is far more sugar than we could ingest in a day on a whole-foods diet.

fruit drinks are often made from fruit juice concentrates (water has been removed from the juice) but frozen fruit juice concentrates that can be reconstituted to make fruit juice often contain added sugar, too. Dairy desserts include ice cream, sweetened yogurt, and sweetened milk. A lot of sugar is often added to coffee drinks. On average, added sugars make up about 14 percent of total dietary energy in the United States, and most added sugar is eaten at home.[9] This gives us an opportunity to have a big impact on our consumption of added sugar, because we have more control over what we consume at home than the food we consume when we're away from home. Many of us were raised to think that a high level of sugar consumption is natural. In reality, it is far from natural! For example, imagine adding nine teaspoons of sugar to a glass of water and gulping it down. Seems like a sickening idea, doesn't it? And yet, a typical twelve-ounce can of soft drink contains nine teaspoons of sugar.

The best solution for obesity due to eating too much sugar is to eat less sugar. Many of us are carrying too much fat because we are consuming too much added sugar. So our first basic guideline for healthier eating is to *eat less added sugar*. One easy way to avoid added sugar is to eat foods only as they are found in nature. If more of us followed this one simple guideline, there would be much less obesity in the world. This guideline has been proposed many times by doctors, scientists, and wise parents, yet it seems to remain a secret that many of us don't know.

To demonstrate how uniquely problematic sugar is as a carbohydrate, consider the recent tenfold increase in diabetes in China. In 1980 the incidence of diabetes in China was 1 percent. By 2013, it had increased to over 11 percent.[10] The increase has been attributed to the industrialization and economic growth of China, which have brought many changes, including decreased physical activity. However, even though the traditional Chinese diet has always been high in carbohydrates, there has been a big increase in sugar consumption since 1980. A 2013 review of the medical research reported that one reason many people can have difficulty controlling their consumption of available foods high in sugar is that the sugar affects the brain and generates a rewarding sensation. They found that this can lead to addictive behavior that can lead to the use of addictive drugs.[11] People I know have reported that they have successfully used fastercise to cancel a sugar craving.

Refined Carbohydrates

Though added sugar is probably the most problematic of all refined carbohydrates, other refined carbohydrates are also profoundly contributing to obesity. Starch is the main carbohydrate in our diets, and most of the starches we eat come from plant-based foods. Just as our bodies store energy in the form of glycogen for later use, plants store energy in the form of starch. Plants make starch by connecting glucose molecules together. A key point here is that, unlike table sugar, the starch we eat is built entirely from glucose.

Plants also contain a lot of structural carbohydrates such as cellulose. Cellulose is one of the materials that a plant uses to build its own form and structure. Think of cellulose as the structural material that the plant uses to make "storage bins" in which to store starch.

As it turns out, our digestive systems can easily break down starch, but not cellulose. Nevertheless, cellulose plays an important role as insoluble fiber that helps us maintain a healthy digestive tract. Termites have micro-organisms in their midguts that can digest cellulose, which is why termites can subsist on eating wood and humans can't. Structural carbohydrates produced by plants are rough and fibrous compared to the smooth and easily digested storage carbohydrates, which is why dietary fiber is also referred to as roughage. It's easy to see how eating plant material as it is found in nature is very different from eating processed and refined carbohydrates. Eating unrefined rice, beans, potatoes, wheat, and corn is like eating the storage bins along with the stored starch. The food industry makes these unrefined foods easier to cook and easier to chew and digest by grinding them with machines or processing them with chemicals. Eating plant foods after this type of processing is like eating cracked storage bins with the storage spilling out. To make these food items even smoother, faster to cook, and easier to digest, the food industry refines them even further by removing the fiber (storage bins) altogether, so we end up eating only the starch. Refining carbohydrates in this way can have explosive conse-quences. Grinding carbohydrates into fine particles results in easier access to the energy they store. If you google "grain explosion" or "dust explosion" you'll find videos of the fireballs that can occur when a tiny spark acciden-tally ignites carbohydrate dust suspended in the air. Remember, our bodies *burn* fuel, only without the flame. In our case, consuming large amounts of refined carbohydrates is resulting in an explosion of obesity.

The *glycemic index* is a measure of how much blood glucose level changes within two hours of eating a food as compared to eating straight glucose. Because starch is built entirely from glucose, something surprising happens when we eat starch without its intact storage bins (that is, when we eat refined carbohydrates). As shown in figure 8.2, many refined carbohydrates cause glucose levels to spike even faster than table sugar does! Boom!

Table sugar, which is heavily refined from sugar beets and sugar cane, has had all its storage bins removed, but because it is only 50 percent glucose, its glycemic index is considerably lower than that of glucose. Notice in the graph that rolled oats (oats that have had some of their storage bins

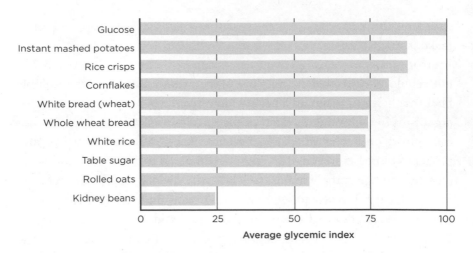

Figure 8.2. This chart, which is based on data from multiple studies by different research labs, shows the glycemic index of many common foods. Glycemic index is a measure of how much a food increases blood sugar levels within two hours, as compared to glucose itself. Source: F. Atkinson, K. Foster-Powell, and J. Brand-Miller, "International Tables of Glycemic Index and Glycemic Load Values: 2008," *Diabetes Care* 31 (2008): 2281–83.

cracked) have a lower glycemic index than table sugar, while unrefined kidney beans (storage bins intact) have the lowest of all.

The glycemic index of white rice appears to be higher than table sugar, but table sugar is more fattening than rice because of the unique effects of fructose on the liver. It's easy to see that *combining* refined carbohydrates with sugar would be especially fattening, and this is precisely what many people eat in excess: breakfast cereal, cookies, cakes, snack bars, pastries, pizza with a soft drink. Animals in the wild are rarely obese, because they eat unrefined food and their food supply is balanced with population size. (The more food the more animals, the less food the fewer animals.) But many people today have easy access to an almost unlimited supply of refined food. Boom! Thus, the second secret hidden in plain sight is also the second guideline for reversing obesity: *Avoid refined and processed foods.*

Refined Fats

Just as eating refined or processed carbohydrates can be problematic, so can eating refined or processed fats. Our bodies are well equipped to digest the foods that we've been eating for tens of thousands of years. However, modern food production techniques have changed our food supply more in the last one hundred years than in the thousands of years before that. For example, to make vegetable oil, plant material is often heated with chemical solvents to extract the oil. Acid may be added to break down solids, and more chemicals may be added to improve color, deodorize the oil, and separate the oil into different components. Then, the oils may be hydrogenated to generate trans bonds in the fats, so that the oil will remain solid at room temperature. (This makes for spreadable products like margarine.) When we consume these foreign trans-fatty acids, they are incorporated into our cellular membranes, leading to disruption of normal functions. Cold-pressed or expeller-pressed oils are preferable because obtaining them doesn't involve high heat. It is a simple pressing process that extracts the oil from the plant material. I feel that organic cold-pressed or expeller-pressed olive oil is a favorable option. Coconut oil is also a favorable option, as are the fats in avocados, nuts, seeds, fish, eggs, meat, and dairy (especially grass-fed, pastured, and organic).

Eat Real to Eat Right

Food as it is found in nature is *real* food. Some examples of real food include unprocessed fruits, vegetables, brown rice and other whole grains, beans, nuts, meat, fish, eggs, and dairy. These unrefined foods curb obesity in a few highly effective ways. These foods still have intact storage bins in the form of fiber. The storage bins take up space. This makes these foods less energy dense so there are fewer calories in a mouthful. The fiber content also makes these foods more filling. There are stretch receptors in our stomachs that stimulate the release of satiety hormones when our stomachs get full. In addition, the storage bins slow the release of the energy contained in the food. Thus, eating real food is another secret for losing fat that many people don't know.

We all love convenience, and not only does the convenience factor discourage foraging behavior but it is perhaps the biggest driver for the consumption of processed and refined foods. How can you distinguish whole foods from processed foods? Processed foods are typically found in the center aisles of a grocery store; unprocessed foods are usually on the perimeter. Read package labels and avoid food products that contain ingredients that sound like something that's the result of chemical processing, such as disodium guanylate.

One kitchen appliance that makes preparing meals based on real food easier and more convenient is an electric pressure cooker. It can cook rice in eight minutes and dry beans in fifty minutes. It can cook potatoes and vegetables in less than ten minutes. You can set the cooker's timer for a delayed start, so that it will be hot and ready when you get home from work or school. So easy. Crock-Pots are another option. How to conveniently and easily switch to a whole-foods diet is beyond the scope of this book, but fortunately, there are an amazing number of great resources readily available. Whenever I want to learn how to do anything I look on YouTube, and within minutes I usually find a video that shows me a way. Some food stores now offer cooking classes as well. My daughters have become gourmet cooks by watching Food Network shows. I feel the key to the diet side of the fastercise program is the same as the exercise side: Just get started. You can adjust as you go along.

Protein

Adequate protein intake can be helpful in preserving muscle, building muscle, burning fat, and increasing satiety. Eating insufficient protein can contribute to increased muscle loss.[12] This is especially true during periods of caloric deficit. Therefore, if your goal is to lose weight, it might be wise to consume *more* protein than you would if your goal is to maintain your current weight or to gain weight. A caveat: People usually need less protein when they are adapted to a low-carb diet (keto-adapted). You may notice that you feel warm after eating a high-protein meal, and that's because it takes more energy to digest protein than it does carbohydrate.[13]

One gram of dietary protein per pound of body weight per day is a standard amount to consume for maintaining or building muscle. The more fat your body contains, the smaller the proportion of your body weight is muscle. Therefore, you may need less than 1 gram of dietary protein per pound of body weight per day in order to maintain your muscle. Conversely, the more muscular you are, the more dietary protein you will need to consume to maintain your muscles. Although sufficient protein intake is important, avoid eating too much, because overeating can signal your body to shift more to storage mode, and excess protein can raise insulin levels and contribute to fat production. As you probably already know, protein is found in foods such as fish, beans, meat, eggs, and dairy. Vegetables can also include protein. And while real food as it is found in nature is always preferable, organic protein powders made from pea, rice, whey, and beef can be a convenient way to increase the quantity of protein in a diet. Nuts are also a good source of protein; note that nuts often contain more than twice as much fat as protein, though. Beans are an inexpensive source of protein and also contain unrefined carbohydrates. They often contain more than twice as much carbohydrate as protein.

Carbohydrate

Carbohydrates are useful as immediate fuel, for replenishing glycogen, and to promote muscle gain and fat loss. However, as noted previously, excess carbohydrates can contribute to obesity, insulin resistance, metabolic syndrome, and diabetes. If you are overweight or obese, you may have some insulin resistance and you may want to reduce your carbohydrate intake (especially sugar and refined carbohydrates). While some starches release glucose during digestion in the small intestine, those known as resistant starches do not. Resistant starches provide fiber that is fermented by the bacteria in the large intestine to produce short-chain fatty acids that feed friendly bacteria in our guts. Resistant starch may increase insulin sensitivity and reduce the risk of type 2 diabetes.[14] Thus, it is important to include some resistant starch in your diet, and it can be found in grains, seeds, legumes, and potatoes and rice that have been cooked and cooled.

Vegetables such as broccoli, cauliflower, green beans, carrots, Brussels sprouts, cabbage, kale, onions, spinach, peppers, leafy greens, tomatoes, peas, and squash are wonderful choices because they contain complex, unrefined carbohydrates. Bananas and plantains have a lot of starch. Fruits such as apples, pears, grapes, and watermelon contain cellulose as well as small amounts of fructose. Fruit can be a fine choice for dessert.

When we overeat carbohydrates, our energy expenditure increases more than when we overeat fats. This explains why we sometimes feel a surge of energy after eating carbs. Eating carbohydrates in the afternoon is helpful for lowering stress, replenishing glycogen, and preparing for a good night's sleep. On the other hand, eating a large quantity of carbohydrates, particularly sweets and refined carbohydrates, may hinder muscle growth overnight due to the countering of growth hormone by insulin.

Fat

It's important to include fats in your diet in a balance that suits your individual genetics and circumstances. Fats are important components of the cell membranes of all the cells in the body, including brain cells. Fats are essential for optimal cell, brain, hormone, and immune function. At the same time, overfeeding fat can lead to greater fat accumulation than overfeeding carbohydrates.[15]

Different types of fat play different roles in the body. For example, the saturated fats found in meat and dairy can be useful in making hormones and cell membranes. Polyunsaturated fats such as omega-6 and omega-3 fatty acids reduce the risk of heart disease, lower "bad" LDL cholesterol, raise "good" HDL cholesterol, and reduce cancer risk. Medical studies have shown improvement in various health conditions when the ratio of omega-6 to omega-3 fatty acids in the diet is kept low (from 1:1 to 4:1 or maybe 5:1).[16] Unfortunately, many of us consume a much higher ratio of omega-6 fatty acids to omega-3 fatty acids, which can lead to weight gain, poor immune function, and chronic inflammation. Omega-6 oils are found in nuts, seeds, and vegetable oils (corn, safflower, soybean, borage, and evening primrose oil). Omega-3 oils are found in fish, eggs, macadamia nuts, walnuts, flax oil, and meat from grass-fed animals. Therefore, many health experts are advocating that we eat more fish and nuts and less processed corn and vegetable oils. Both omega-6 and omega-3 oils are beneficial, but the diets of our foraging ancestors had a much lower ratio of omega-6 to omega-3 oils than is often found in our modern Western diet. A balance of olive oil, animal fat, and coconut oil seems like a reasonable choice.

Choose oils that haven't been altered by chemical processing or extreme heat. As mentioned earlier, expeller- or cold-pressed olive oil is one healthy option. Coconut oil is another. Coconut oil contains both medium-chain and long-chain triglycerides. By definition, medium-chain triglycerides (MCTs) are six to twelve carbons in length. Unlike long-chain fatty acids, the medium-chain fatty acids can enter cells without the assistance of cell membrane fatty acid transport proteins. In addition, they don't require cytosolic transport proteins and can enter the mitochondria for oxidation

Avoiding Toxins

Our world has become so polluted that we are all exposed to toxins. There are toxins in the air that end up in the rain that falls on crop fields and plants. Many toxins are fat soluble, so when humans and other animals eat the plants, the toxins tend to become concentrated in their body fat. This can be true even for animals that have been raised organically (without the intentional use of pesticides and other chemicals). However, the fats from animals that have been raised organically may contain fewer toxins, and studies show that organically raised produce contains fewer pesticide residues than conventionally raised produce. Determining food quality can be a complex issue. There can be unexpected toxicity problems related to diet, such as the account of a man whose blood mercury levels soared after he embarked on a seafood diet. My point is not to frighten or frustrate you, but to make you aware that it may not be easy to find uncontaminated food. I recommend that you do the best you can to search out (which can take time) and buy the freshest and least contaminated food available to you.

without the assistance of the carnitine shuttle. Because of this, oxidation of MCTs is about five times faster than oxidation of long-chain fatty acids.[17] Also MCTs are less likely than long-chain triglycerides to be stored as fat. As compared to long-chain triglycerides, MCTs have been found to lower LDL cholesterol and body weight and increase HDL in mice with type 2 diabetes.[18] MCTs may also be helpful in preventing Alzheimer's disease.[19] MCTs can be isolated from coconut oil and sold as a separate product.

Trans-fatty acids are found naturally in trace amounts in some foods (3 percent in butter). The word *trans* refers to the shape of the fatty acid molecule. In nature, most fatty acids are in the *cis* shape, not the trans shape. However, once the food industry discovered that trans fats were more spreadable and less expensive than other oils for baking, production of products containing trans fats—and thus consumption of trans fats— rose dramatically. Now, a recognized body of research has shown that trans fats raise LDL, lower HDL, and increase the risk of cardiovascular disease and the development of type 2 diabetes. Scientists agree that the consumption of trans fats should be reduced again to trace amounts, and in May of 2018 the World Health Organization released REPLACE, a step-by-step guide for the elimination of industrially produced trans-fatty acids from the global food supply.

Vitamins

One reason why vitamin intake matters is that many vitamins play an essential role in mitochondrial function and energy production. For example, thiamine (B_1) helps convert pyruvate into acetyl-CoA; riboflavin (B_2) helps shuttle electrons from the citric acid cycle (CAC) to oxidative phosphorylation (OXPHOS); niacin (B_3) is a precursor of NADH; pantothenic acid (B_5) is a precursor of coenzyme A, which is used to make acetyl-CoA; pyridoxine (B_6) helps enzymes catalyze energy reactions and is also important for making epinephrine; cobalamin (B_{12}) supports the production of phosphocreatine as well as parts of the electron transport chain; alpha-lipoic acid aids NADH electron transport and reduces reactive oxygen species; and coenzyme Q_{10} is part of the electron transport chain.[20] Vitamin A regulates how much pyruvate enters the CAC. Vitamin E prevents mitochondrial damage. Pyrroloquinoline quinone (PQQ) increases the size and number of mitochondria. All of these vitamins are provided by a healthy diet, and you can take a high-quality multivitamin supplement daily to ensure adequate levels.

Figuring Out What Works for You

Simply following the three basic guidelines of avoiding added sugars, avoiding refined foods, and eating real food will help many of us have a lot less trouble losing fat. Beyond that, the next step is to consider the proportions of proteins, carbohydrates, and fats in your diet. Discussions of this subject can be a little like discussing politics or religion. Two of us can be equally well informed, convinced, persuasive, passionate, and sometimes dogmatic on the topic even though we have seemingly conflicting views.

Maybe it's because we're looking at things from different perspectives. Maybe we're both right, maybe neither of us is right, maybe it depends on the person and circumstances. Each one of us ultimately needs to figure out what helps us accomplish our goals. One thing I've learned is that what we know for *certain* to be true today may turn out not to be so true next month, or next year. So let's consider the following thoughts about food as simply food for thought. Overall, the goal is to eat healthy, live healthy, and be healthy in a comfortable, convenient, sensible, and sustainable way.

There's Not Just One

As I said at the start of this chapter, I don't believe there is just one healthy diet everyone needs to eat in order to achieve optimal fitness. It all depends on the person, their circumstances, and what they're trying to accomplish. To me, diets are like trucks. There are fire trucks, dump trucks, garbage trucks, cement trucks, pickup trucks, on so on. They're all useful for hauling stuff, although a cement truck would be not be a good substitute if someone needed a fire truck, and a fire truck would not haul liquid cement very well. However, each truck is perfectly suited for hauling some type of material. Similarly, a recent study showed that each of four different diets worked just fine at helping people lose weight. This *New England Journal of Medicine* study involving more than eight hundred patients compared the weight-loss results for four diets that had different compositions of fat, protein, and carbohydrates.[21] The research was conducted by Harvard University and Louisiana State University researchers who tested two low-fat diets, one that had a higher percentage of protein and one with a lower percentage. They also tested two high-fat diets, and again, one had a higher percentage of protein and the other a lower percentage. The authors concluded that after two years, the weight-loss results were similar regardless of the type of diet followed.

Diet and exercise are tools that we can use in many different ways to accomplish many different things. For example, low-carb diets have been used to address epilepsy and Alzheimer's disease.[22]

Another famous weight-loss experiment was conducted by Kansas State University nutrition professor Mark Haub. For ten weeks, Haub followed a diet that consisted of one-third protein, vegetables, and a multivitamin. The other two-thirds consisted of Twinkies, Nutty Bars, powdered sugar doughnuts, other Hostess and Little Debbie snacks, Doritos, Oreos, and sugary cereals. On this diet, Haub was able to lose twenty-seven pounds over the course of the ten weeks. He said his success boiled down mostly to portion control. With the loss of weight, his LDL or "bad" cholesterol dropped 20 percent and his HDL or "good" cholesterol increased by 20 percent and his triglycerides dropped by 39 percent. Haub did *not* conclude that this was a health-promoting diet, but he was able to lose weight.

Why have I told you about these research studies? My purpose is to demonstrate that *portion control* is probably a more important aspect of a diet than the proportion of fat, protein, and carbohydrates in your diet. And when it comes to portion control, hunger is our friend. Really? Absolutely. Your sense of hunger is like your super smart, automated personal assistant. Calculating and counting the number of calories we need to ingest as protein, carbohydrate, and fat can be very time consuming and mind-boggling in complexity. The calculations are based on theories that may or may not be sound, and calorie charts aren't always accurate. Fortunately, our bodies can handle all the complexities for us. And as I've explained throughout this book, the sensation of hunger lets us know just when the energy supply from our most recent meal is running out and when we're starting to burn muscle. We can simply pay attention and use hunger as our guide. Eat real food, and during a meal, be mindful of your hunger and stop eating when you feel satisfied. Use fastercise to cancel your hunger pangs so that you replace some of your eating with snacking on your fat storage. Fastercise immediately before eating, as we discussed in the previous chapter and then eat mindfully—stopping when you're full. You may not be able to eat whatever you want whenever you want, but you are free to eat whatever you need whenever you need in order to feel very comfortable getting great results. Not only is it possible to lose weight without suffering from hunger pangs, for many reasons it may be the best way to do so.

Occasionally you may choose to eat not from hunger but because it's fun or for comfort, and there's nothing wrong with that, in moderation. Certainly, if we eat healthy even 80 percent of the time we can still get good results, keeping in mind that portion control remains important, even when you splurge on your favorite flavor of ice cream.

In the next chapter, I describe some differences between a high-carb diet and a low-carb diet and how to most effectively integrate fastercise with either type of diet. I discuss how to vary the percentages of dietary protein, carbohydrate, and fat to suit your individual needs and circumstances. As mentioned earlier, your body has the capacity to undergo significant adaptations in response to diet and exercise. If you want to know how many grams of protein, carbohydrates, or fat are in a given food, you can read the product label. Or if it's real food that doesn't have a label, you can simply google: "macronutrients in _____," filling in the blank with the name of that food. Remember that individual food ingredients typically contain some combination of protein, carbohydrate, and fat, and each of these types of macronutrients has benefits for your body.

Keeping Your Perspective

Everyone has the occasional wish to enjoy a big meal, such as during a holiday family dinner or when going out to dinner at a special restaurant. One of the beauties of fastercise is that it helps you handle these infrequent binges. In chapter 6, I explained how to strategically use shivercise to deplete your glycogen stores before a feast. You can also use shivercise after the feast to deplete your glycogen before your next meal. The general idea is to balance a big-eating meal by spending more time before the meal in the non-fed state, and then again after you've swung through the fed state following a feast. For example, after a big holiday dinner, don't go to the kitchen for a helping of leftovers a few hours later. Yes, they're tempting, and you might even think you're hungry again, but you're not! You can fastercise instead, and go to bed. The leftovers will still be there the following day. You can eat delicious food after you've gotten hungry and fastercised again.

How about those times when you want desperately to eat something sweet? Fastercise can help. You can also try something with a little fat in it, like a spoonful of peanut butter. Another good option is to satisfy the craving with a healthy choice—fresh fruit. Certain bitter herbs such as dandelion, burdock root, and orange peel can also reduce sweet cravings. Sometimes, we may want something warm and relaxing. Consider some warm herbal tea. It's delicious and filling and satisfying and many cultures serve tea after dinner as opposed to dessert. If you're determined to eat something sugary, then choose a sweet treat that contains complex carbohydrates instead of refined carbohydrates. It may sound strange but potatoes, beans, or oatmeal topped with some powdered sugar or honey would be a better choice than a sweet food full of refined carbohydrates such as cookies, cake, or doughnuts. I've tried sugar on boiled potatoes or beans and it's quite tasty—I'm surprised it's not more popular. But let's say you are determined to eat a refined carbohydrate dessert. It is more sensible to do so when your insulin levels are already high after eating. That way, at least you don't interrupt your time in the non-fed state, which you will if you snack on a cookie or drink a sweetened beverage when you are hungry between meals (and you could fastercise instead).

When we eat fewer meals a day, we can use the time, energy, and money we save on the skipped meal to seek out delicious and enjoyable new recipes, learn new cooking skills, use fastercise to heighten our sensitivity to and appreciation for food, and then relish the reward of a well-earned meal. Instead of our trudging through food preparation and eating less appetizing meals, each meal will be a nourishing and special treat.

CHAPTER 9

Managing Diets
and Fastercise

D ifferent kinds of diets are like different songs played on a piano.
No matter what the song, you use the same ten fingers, and you
have the same range of notes. But depending on which piano
keys you press and how, the songs can sound quite different. Different
diets all draw from the same potential range of foods and our bodies use
the same chemical processes to process that food. In this chapter, the focus
is how to choose between a high-carb approach or a low-carb approach to
combining diet and fastercise and how to integrate fastercise techniques
(shivercise and tightercise) with high-carb diets and low-carb diets. Both
of these diets are "songs" that our bodies can learn to play beautifully. Both
have their challenges, and both have their appeal.

High-Carb Versus Low-Carb

There's been a lot of discussion in the media about high-carb and low-carb
diets, and people want to know what balance of macronutrients to eat. In
a way, the answer to this question boils down to deciding where you want
to be on the high-carb-to-low-carb spectrum.

Have you ever noticed that you don't hear much about low-protein
diets? That's because a low-protein diet is usually prescribed only for people
with inherited metabolic, kidney, or liver disorders who have a difficult
time processing protein. Otherwise, we all benefit from a certain amount
of protein. For that reason, of the three macronutrients—carbohydrates,

protein, and fat—protein is fairly well pinned down. We first decide what level of protein we want in our diet and then we go from there.

The two remaining macronutrients left to balance are carbohydrates and fats. However, as we eat a higher percentage of calories from one of these macronutrients, we automatically eat a lower percentage of calories from the other. Thus, a high-carb diet is a low-fat diet and a low-carb diet is a high-fat diet. The answer to the question of what balance of macronutrients to eat boils down to a discussion of high-carb versus low-carb.

There is a lot of controversy about ketogenic (low-carb) diets and ketosis. There are strong critics of ketogenic diets who say they can be harmful and unsustainable. After hundreds of hours reviewing research about low-carb diets, I have learned that we can't be overly certain about any of our opinions. Our opinions are typically based on our own experience, the experience of others, and on research studies that often contradict one another. It's very easy to jump to erroneous or oversimplified conclusions. And then there's ignorance. As a physician, I didn't know much about low-carb-adaptation until I studied it and experienced it. I'm certain that many physicians aren't very familiar with ketogenic diets or their effects. I suspect that many of your friends and neighbors don't know much about ketogenic diets either.

What Is a High-Carb Diet?

Typically, a high-carb diet consists of approximately 10–40 percent of calories from protein, 40–65 percent from carbohydrates, and 20–35 percent from fats. This diet encompasses a very wide variety of foods and is easy to follow. With a high-carb approach, we can eat plenty of carbohydrates, such as pizza, potatoes, rice, fruit, cereals, oatmeal, and even some dessert. When we are eating with our friends, co-workers, or family members, we can take part in almost everything others are eating.

A reasonable starting point for a high-carb diet is roughly 40 percent of calories from protein, 40 percent from carbohydrate, and 20 percent from fat. Keep in mind that fat contains more calories per gram than protein or carbohydrate, so another way to define these proportions is roughly 45 percent of *food weight* as protein, 45 percent as carbohydrate, and 10 percent as fat. This is a rough estimate, because different foods have different water content. Nevertheless, it does show that on a weight basis, a high-carb diet consists mostly of protein and complex carbs.

What Is a Low-Carb Diet?

Most people in the United States don't eat a ketogenic diet. This can make following a low-carb diet a little less convenient. You may find yourself eating a little differently than your friends and family and having fewer options to choose from at restaurants. Usually, however, you can eat what everyone else is eating, only in different proportions. Eating fewer carbohydrates can be challenging because eating carbohydrates can be pleasurable, especially when your body is carb-adapted. However, the more low-carb-adapted you become, the less quickly your glycogen runs out and the less you crave carbohydrates. Food preferences are largely trained adaptations. Even though following a low-carb diet can take some getting used to, many people feel the benefits are well worth the challenges. How a person implements a ketogenic diet can greatly affect how well they respond to it. Everyone is different and not everyone will do well on a ketogenic diet, but many do.

What constitutes a low-carb diet? First of all, on a low-carb diet we want to obtain most of our calories from fat (stored fat and dietary fat). This will help us build up our fat-burning machinery. Ketogenic diets range from about 60 percent to 90 percent of calories from fat. The more calories from fat, the farther into ketosis we will go.

Of course, everything I discussed in chapter 8 regarding eating real, unprocessed food applies here. Avoid processed meats. Real meat, cheese, eggs, fish, and vegan protein are fine (unless you have a sensitivity or don't function well on one or more of these). Many people think of low-carb diets as eating mostly meat and cheese. However, it is possible for people to stay in ketosis without eating meat.

One approach is to first calculate how much protein you want to consume. A reasonable starting point is 0.5 gram of protein per pound of body weight per day. People who want to grow more muscle, such as children and bodybuilders, might try 0.6–1 gram per pound per day. For the rest of the diet, select greens to supply electrolytes (not to exceed 20–50 grams per day of carbohydrates) along with a liberal mixture of fats. It's very important to take in protein, fats, electrolytes, and water, but it is not essential to eat any carbohydrates, from the standpoint of maintaining the body's tissues. As a reminder, fat contains nine calories per gram while carbohydrates and proteins both contain four calories per gram.

Normal and *popular* are not the same thing. Many people consider a high-carb diet the normal way to eat but low-carb diets are becoming a more popular way to eat. To emphasize just how normal a low-carb diet can be for our bodies, let's consider a few facts. At rest, muscles use primarily fatty acids for fuel. At any given moment, most of the energy circulating in our bloodstream is in the form of fatty acids, not carbohydrates.[1] Also, our bodies normally store twenty times more energy in the form of fat than they do in the form of carbohydrate. It's apparent that it's perfectly normal for our bodies to burn fat for fuel. It bears pointing out that dietary ketosis is a normal condition that occurs with normal fasting or ketogenic diets. Ketosis is different from the much more severe uncontrolled form called ketoacidosis that people with type 1 diabetes can develop when they don't get enough insulin.

The low-carb approach is aimed at helping increase the body's capacity to burn fat, and it leads to higher concentrations of ketones in the blood. Ketosis occurs when the blood level of the ketone beta-hydroxybutyrate reaches 0.5 millimolar (mM) or higher. We are effectively following the low-carb approach when our blood ketone levels are between 0.5 mM and 3.0 mM. Three millimolar is the level at which the muscle-building effects of ketones tend to peak.

It is usually possible to maintain blood ketone concentration above 0.5 mM by restricting carbohydrates to 20–50 grams per day. The specific amount varies from person to person and some can maintain 0.5 mM while consuming as much as 100 grams of carbohydrates per day. That's why measuring blood ketones is so handy. You can ingest carbohydrates from a combination of vegetables, protein, nuts and seeds, fruits, and dairy. Because ketones spare protein, when we're in ketosis we can generally eat only half as much protein as most people eat and still gain muscle.

Low-carb diets ought to be fairly low in protein as well because protein can be converted to glucose, which can lead to lower ketone levels, lessening the benefits that ketones provide. That leaves fat to make up the remainder of the diet. Fat has virtually no effect on insulin levels. Eating a meal of low carbohydrates, moderate protein, and some fat will not drive up insulin in the way a high-carb meal does.

There are some challenges in adapting to a low-carb approach. Insulin levels tend to drop and the body tends to retain less water, resulting in

Measuring Ketone Levels

You can measure your blood ketone levels with blood ketone testing meters available for about $60 (including a starting supply of testing strips). Refill testing strips are available for between $1 and $3 each (depending on the meter). Measuring your blood ketones with such a meter involves poking your finger with a spring-loaded lancet designed to penetrate the skin to the perfect depth. It's surprising, but using the lancet isn't a very painful experience. Dipsticks for measuring ketone levels in urine are available. Though these are less expensive, they can sometimes show negative results even when ketones are present in the blood. Breath analyzers are also available but are more expensive than the blood meter. I prefer using a blood ketone meter.

more frequent urination. This can lead to loss of electrolytes like sodium, potassium, and magnesium, and the development of symptoms such as dehydration, fatigue, weakness, headaches, muscle cramps, palpitations, and muscle twitching. These symptoms are sometimes referred to as the keto flu. Low potassium levels can make gout worse, but taking a potassium supplement (100 milligrams per day) is usually enough to prevent this. In addition, when magnesium and calcium levels are low, kidney stones tend to form more easily in people who are prone to developing them. Those who are prone to kidney stones can increase their intake of dietary calcium, which binds with oxalates in the gut so the oxalates don't form stones as easily in the kidneys. Frequency of urination lessens over time as the kidneys become more low-carb-adapted.

Maintaining electrolyte levels is critical to preventing keto flu on a low-carb diet. When electrolytes get very low people may also notice variations in their heartbeat and rhythm. Susceptible people may even have an increased chance of atrial fibrillation.[2] Supplementing electrolytes and taurine can be very helpful in minimizing such difficulties. One easy way for us to get

Benefits of the Low-Carb Approach

Hunger tends to come on less dramatically.

Burns fat with less risk of weight regain.

May help burn stubborn fat.

Potential to preserve and possibly build muscle even when insulin levels are low.

Improved mitochondrial function, resulting in better endurance.

Reduction in reactive oxygen species, resulting in less oxidative stress.

Reduction in signs of aging and increased longevity.

Potential to reduce the risk of developing Alzheimer's disease.

the extra salt we need is to have an extra 1–2 grams of sodium a day, which can take the form of one or two bouillon cubes.[3] Also, especially during the transition to becoming low-carb-adapted, taking a little extra salt before bedtime can decrease how often we need to get up to urinate.

There are positive effects, too, when we transition from a high-carb diet to a low-carb diet. Because we are transitioning some of our energy dependence from a more quickly depleted energy resource (glycogen) to a more slowly depleted one (fat), our energy levels and moods tend to become more stable. There is less tendency to crave carbs and our appetite can become less demanding. In addition, we can notice big improvements in our short-term memory and enjoy better mental focus, clarity, and learning.

Low-carb diets can reduce the risk of diabetes, high blood pressure, aches and pains, and skin inflammation. Some people may notice that coffee and alcohol tend to have greater effect. Some people temporarily experience bad breath (especially due to exhaled acetone) as they transition to low-carb.

Is a ketogenic diet helpful or harmful for gut health? More research is needed in this area. Cells of the colon love butyrate. (The gut turns fiber into short-chain fatty acids like butyrate.) This is one reason that it's beneficial to include fiber in the diet.

One common mistake people make on a ketogenic diet is that they don't consume enough fat. It can be quite an adjustment after years of being told by the government, doctors, teachers, and parents that we shouldn't eat fat. Eating a high-fat diet can be delicious and satisfying. It's important to not only cut down on carbohydrates and eat protein in moderation but also take in adequate amounts of fat in order to help build up our fat-burning machinery. One way to cut down on carbohydrates is to have "micro desserts." Limit yourself to one mouthful, and savor it fully for the enjoyable flavors. It might be one mouthful of crackers, bread, pie, cake, or ice cream, and we can dress it up with jelly, syrup, or sprinkles since it will be just one bite. That way we can still enjoy the taste without driving our insulin up very much.

When we eat fewer carbohydrates we need to eat more fats in order to maintain our weight. However, if we want to lose weight, we can eat a little less fat than we'd need to maintain it. When people eat a lot of fat they can sometimes experience nausea (especially if they eat too much of the oils rich in omega-6 fatty acids) and diarrhea. When we are losing weight by fasting or with a ketogenic diet and are eating a palatable combination of oils, we tend to feel fine.

For many people, the only thing better than losing excess fat is keeping it off. A 2008 study in the *American Journal of Clinical Nutrition* suggested that regaining fat that we've lost might not be due to returning to our old eating habits as much as it is to hormonal changes that encourage us to eat.[4] This study also demonstrated that one of these changes was less likely on a low-carb diet. Levels of the hormone cholecystokinin (CCK) increase when we eat. This hormone helps us digest fat and protein, and it also suppresses hunger. The study showed that after rapid weight loss of about 10 percent of body weight, CCK levels remained depressed, which stimulated the body to eat. However, as long as the participants were in ketosis, their CCK levels remained the same as before the weight loss.

By inducing hormone-sensitive lipase and lowering insulin levels, low-carb diets help the body to rely more on fat for fuel both at rest and during exercise.[5] During the first few days on a low-carb diet, our muscles use both ketones and fatty acids for fuel. After a couple of weeks, the muscles use mainly fatty acids both at rest and during exercise, which greatly spares ketones for use as a brain fuel, and ketone levels rise significantly.[6] This is part of the reason it can take weeks, and even several months to fully adapt to a low-carb diet.

What's the Right Approach for You?

The differences between a high-carb and a low-carb diet are not drastic; they are more a matter of emphasis, proportions, and adaptations. As we consider the differences, it's helpful to bear in mind their similarities. We always have a combination of carbohydrates, proteins, and fats in our bloodstreams and we are always using a combination of those as fuel. Our brains always require and use some glucose. We always have ketones in our bodies. We burn some fat for fuel whether we are on a high-carb or low-carb diet. All the fuels are present and all the machinery for energy production is there.

A low-carb approach is not a no-carb approach. People on a low-carb diet still eat 20–50 grams of carbohydrates per day. (And even if you were to eat no carbohydrates, your body would still recycle them at the cellular level and transform some protein into carbohydrates.) Those carbohydrates help fuel the brain, the citric acid cycle (CAC), and the replenishing of glycogen. But when you reduce your carbohydrate intake, your body needs to rely more heavily on the breakdown of fat for energy. When we switch from a high-carb approach to a low-carb approach we increase our reliance on fat for fuel and decrease our reliance on carbohydrates. Our bodies adapt to these changes over days and weeks.

Evaluating your current body condition and your goals can help you decide whether to use a high-carb, low-carb, or combination approach. Are you underweight, normal weight, or overweight? What is your primary goal: losing fat or gaining muscle? And are you looking for fast results, or is it more important to choose an eating approach that is easy to implement? To help you answer these questions, let's look at some of the research on glycogen, because as you have learned in previous chapters, glycogen is central to the body's production and use of energy.

Making the Most of Glycogen

Keep in mind that our bodies adapt to the food we eat and the exercise we do. It can take weeks to become fully adapted to a low-carb diet (to become low-carb-adapted). I'll use the term *high-carb-adapted* to describe people who are adapted to a high-carb diet.

In a study of six men who were each randomly assigned to a low-carb diet or high-carb diet, those who had higher glycogen stores (assigned to the high-carb diet) showed an increase in muscle building and reduced muscle loss.[7] The more we can spare glycogen, the more we can preserve and build our muscles. One way to spare glycogen is to eat protein so that the body can burn the protein as fuel. The more dietary protein that is burned as fuel, the more the body's glycogen stores are spared.[8] In addition, protein provides amino acids for muscle building, and a high-protein diet promotes lean muscle mass.[9] Thus, both dietary protein and dietary carbohydrate spare and promote muscle.[10] I consider 1 gram of dietary protein per pound of body weight per day (or higher) to be a high-protein diet because that's enough to promote muscle gain and probably enough to prevent ketosis. To my way of thinking a high-protein diet produces many of the same effects as a high-carb diet. A low-carb diet can lead the body to ketosis, but a high-protein diet can't. Such a large amount of the dietary protein is converted to carbohydrates that ketosis doesn't develop. In effect, a high-protein diet is a high-carb diet.

Dietary protein and fiber have been shown to independently reduce hunger.[11] This supports the idea that hunger is a sign we are burning muscle, considering a high-protein diet would tend to spare and build muscle, thereby reducing hunger. The more muscle we have, the more mitochondria we have, which means our bodies have greater capacity to burn fat. This might help explain why people who are on a high-protein diet tend to burn more abdominal fat.[12]

Our glycogen levels largely determine whether we are high-carb-adapted or low-carb-adapted. It takes days of carbohydrate overfeeding to completely max out the body's glycogen stores, so most people do not have completely filled glycogen storage. The body can store about four times more glycogen in muscle tissue than in the liver. The more muscle a person has the more glycogen they can store; a human's full-body glycogen stores can vary from about 500 milligrams to 1.1 kilogram.[13] The higher our glycogen stores, the more we rely on glucose for fuel; the lower our glycogen stores, the more we rely on fat for fuel.[14] When glycogen levels are very low, we can eat a lot of carbohydrates without building much fat

Water Weight

You'll recall that the body stores a lot of water with glycogen: three to four pounds of water for every pound of glycogen. That can lead to some deceptive weight-loss results. You can deplete your glycogen by not eating much for a day, or by jogging for an hour or two. You'll weigh yourself and find that you've lost three pounds, and it will be very exciting! But this drop in pounds does not represent fat loss. When you eat again after your fast or jogging session, your insulin will rise, which will signal your kidneys to retain more water and your liver and muscles to make more glycogen. And you'll probably regain those three pounds. So, it's best not to judge your success or failure in fat loss by your day-to-day weight readings, because weight is not a reliable measurement of body fat!

because glycogen stores will be replenished first.[15] Glycogen levels need to fill substantially before the body begins to make and store fat.[16]

The body has the greatest potential to store glycogen after glycogen has been depleted completely. (This sensitizes the body to gain glycogen.)[17] Glycogen can be completely depleted by one to two hours of cardio exercise.[18] Insulin-independent rapid glycogen replenishment can occur within thirty to sixty minutes after glycogen-depleting exercise. (Slower, insulin-dependent replenishment occurs for hours after that.)[19] Glycogen can be replenished from lactate and glycerol as well as from food. With effort, glycogen stores can be largely replenished in one day.

When you shivercise, you'll find your appetite is eliminated after a minute or two. Our hunger tends to go away faster and stays gone longer after shivercise when our glycogen levels are full. Occasionally, our hunger might seem to get a little more intense for ten or twenty seconds after we shivercise (because of the glucose we've used up) before it goes away in a minute or two because of the stored fuel we've liberated. When our glycogen levels are full and we cancel hunger with shivercise, our hunger tends to remain gone for one and a half to two hours. When our hunger comes back we can shivercise again. As our glycogen gets depleted our hunger tends to come back sooner, within twenty to thirty minutes. Glycogen tends to get depleted faster and hunger tends to come back more strongly when we are using the high-carb approach. When we are low-carb-adapted, our bodies rely less heavily on glycogen.

When we rely more on fat for fuel, our bodies can use some of that fuel to power the recycling of carbohydrates back into stored glycogen. Thus, even though we're not eating a lot of new carbohydrates, our body can spare and recycle a lot of our existing carbohydrates to rebuild glycogen.[20]

Changing Body Composition

There are several ways you can change your body composition. You can lose fat, build muscle, or both. If you want to change your body composition, which would be better for you, following a high-carb diet or a low-carb diet? That may depend on whether you care more about losing fat, gaining muscle, or both. Both insulin and ketones have protein-sparing and muscle-building effects. In the high-carb approach, we rely fairly heavily on the effects of exercise and *insulin* to help build muscles. In the low-carb approach, we rely more on the effects of exercise and *ketones* to help build muscles. One study involving eleven women and sixteen men showed that a high-carb diet plus resistance exercise resulted in more muscle gain (1.8 kilograms versus 1.0 kilogram) but less fat loss (–3.5 kilograms versus –7.7 kilograms) than a low-carb diet plus resistance exercise.[21] It makes sense that a high-carb diet would build more muscle because insulin may spare muscle more strongly than ketones do. It also makes sense that a high-carb diet would result in about half as much fat loss because insulin promotes *fat storing* while ketones result from *fat burning*. If we compare the change in body composition with the two approaches by adding the kilograms of muscle gained *plus* the kilograms of fat lost for each group, we get a total change of 5.3 kilograms for the high-carb group and 8.7 kilograms for the low-carb group. The low-carb group made 64 percent more progress in losing fat even while gaining muscle. Thus, if your primary goal is building muscle and losing fat is secondary, you might try the high-carb approach. If losing fat is the primary goal and building muscle is secondary, you might find success through the low-carb approach.

Some people may find it helpful to alternate between a high-carb diet and a low-carb diet from time to time. However, keep in mind that when you're following the low-carb approach, taking a break and eating a lot of carbs during a "cheat meal" can reduce your ability to burn fat for days, not just hours.[22] Nevertheless, both approaches are available and if you wish to use one approach for days, weeks, or months and then switch to the other one, you can certainly benefit from either. Once you gain experience with these approaches you can implement them to suit your preferences, lifestyles, and goals.

Diet Effects on the CAC

In order to understand how to integrate fastercise most effectively with diet, we first need to understand more details about the chemical pathway that supplies most of our body's energy, the citric acid cycle (CAC, or Krebs cycle). We began discussing the CAC in chapter 2.

The CAC takes place in the infinitely complicated soupy conditions of the mitochondria, but we can conceptualize it with more structure to make it easier to understand. Let's liken the CAC to a machine that uses fuel to produce energy. The chemical reactions and intermediates of the cycle are the machine. What's a little different about this machine is that some of the same materials that contribute to the *construction* of the machine also contribute to *fueling* the machine.

For our purposes, we can think of acetyl-CoA as the fuel of the CAC. But in the bigger picture, the source of acetyl-CoA is fats, proteins, and carbohydrates. All three of these categories of macronutrient can supply fuel to the CAC. And both carbohydrates and protein can also contribute to the intermediates of the CAC, as shown in figure 9.1. Fats do not contribute intermediates directly, but ketones, which are derived from fats, can help speed up the chemical reaction that converts one of the intermediates (succinyl-CoA) into another (succinate). Ketones can also be converted into acetyl-CoA and can have an influence on the CAC in other ways, which I describe later in this chapter.

During each turn of the CAC, two carbon atoms (in the form of acetyl-CoA) enter the cycle and two carbon atoms (in the form of CO_2) leave the cycle. In the process, the CAC transfers energy from the fuel

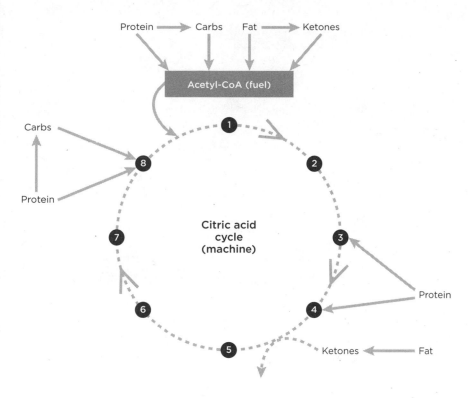

Figure 9.1. Protein, carbohydrates, and fats all feed into the citric acid cycle (CAC) as precursors of acetyl-CoA. Protein and carbohydrates can also supply intermediates.

into high-energy molecules called NADH and FADH$_2$. During oxidative phosphorylation (OXPHOS), energy from NADH and FADH$_2$ is transferred into ATP. This process is how we generate the vast majority of the energy we need to operate and build, repair, recharge, and protect our bodies.

Just as the thermostat in a house signals the heating system to shut off when the house has warmed up, the CAC has regulatory mechanisms that cause it to slow down when there is sufficient NADH and ATP. When we fastercise, our bodies consume ATP faster than OXPHOS can produce it, and ATP levels can decline even though NADH levels are still relatively high. When NADH is high, the first, third, and fourth reactions of the CAC are inhibited. However, when ATP levels are low, the first

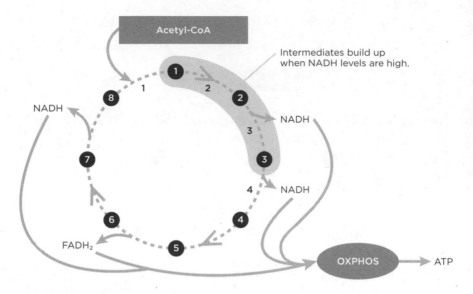

Figure 9.2. As long as NADH levels are high, the first, third, and fourth reactions of the CAC are inhibited. When ATP levels drop, the first and third reactions speed up. When NADH is high and ATP is low, there is a buildup in the intermediates of the first third of the CAC.

and third reactions of the CAC are stimulated. This allows a buildup of the intermediates in the first third of the CAC that stand ready to supply OXPHOS with as much NADH as it can handle. The first three reactions are proceeding, but the fourth reaction is still bottlenecked due to the high NADH level, as shown in figure 9.2. When ATP is being used up quickly, it's important for the OXPHOS process to be kept fully supplied with NADH and $FADH_2$ so it can produce as much ATP as possible.

As soon as the NADH level begins to decrease, NADH can be instantly replenished from the buildup of intermediates readily available to supply the third and fourth reactions (which are the reactions that produce NADH). In other words, our ability to burn fat, generate ATP, and use ATP to build muscle is limited by how much OXPHOS capacity we have, which is determined by the number and size of our mitochondria. George Brooks at the University of California has performed research showing that lactate increases mitochondrial biosynthesis in a variety of ways.[23]

Since fastercise leads the body to produce lactate, we can reliably use fastercise to increase the size and number of our mitochondria and thus to increase our OXPHOS capacity.

Ketones and the CAC

Our bodies make ketones all the time, no matter what diet we're on. However, ketone levels start to rise whenever carbohydrate consumption drops, whether because of our food choices or due to fasting. Ketones are burned for fuel in the mitochondria of all bodily tissues except the liver. A combination of three conditions signals the body to increase the production of ketones: low insulin, high fatty acids, and low glycerol.[24] When carbohydrate consumption drops and glucose levels drop, insulin levels drop. The body breaks down fat to release fatty acids and to increase the enzymes that make ketones. Low glycerol is a sign that carbohydrate levels are low and that more ketones are needed to help support brain function.

It is harder for us to store fat when we are in ketosis, because our carbohydrate supply is low. When our carbohydrate supply is low, our glycerol levels are low, and glycerol is a component of triglycerides. However, it *is* possible to gain fat on a ketogenic diet if we overeat, because excess dietary fat can be stored in our fat stores.

Ketones and carbohydrates can both facilitate fat burning in the mitochondria. Ketones actually contain more energy than pyruvate, and ketones can deliver 28 percent more power than pyruvate can for the generation of ATP.[25] When the body is low-carb-adapted, it increases its reliance on ketones, and when it is high-carb-adapted, it increases its reliance on carbohydrates to serve that purpose. In the high-carb approach, we can use shivercise to mobilize just enough glycogen to provide the carbohydrates necessary to maximize fat burning. If we don't have enough carbohydrates our fat-burning will slow down, but if we have more than we need (because we ate or drank something caloric), then our insulin might go up and our fat burning could stop. We might even make and store some fat. In the low-carb approach, it appears that our bodies generate just enough ketones from fat to maximize fat burning. A ketogenic diet can raise our ketone levels 2–6 mM. Taking exogenous ketones might increase our ketone levels (by 1 mM at the most), but they might temporarily reduce our fat burning since they can also be used as fuel and spare some fat.

High-Carb Approach and the CAC

Let's consider the effect on the CAC when we eat a large meal on a high-carb diet. The short answer is that a lot begins to happen, as shown in figure 9.3. Some of the protein, carbohydrates, and fat from the food are converted to acetyl-CoA and become fuel for the CAC. At the same time, some of the carbohydrates and protein are converted into CAC intermediates (especially oxaloacetate). You'll recall that adding intermediates to the CAC is called anaplerosis, and it's like adding extra lanes to the CAC energy roundabout. With all of this fuel flooding in, once our bodies generate sufficient NADH to power ATP production through OXPHOS to meet our current needs, the presence of the NADH starts to impede some of the CAC reactions, and the intermediates of the first third of the CAC can build up. Excess energy can be stored as fat. In addition, excess

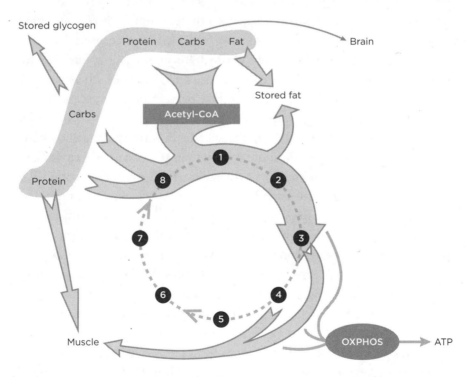

Figure 9.3. An influx of protein, carbohydrates, and fat from a large meal provides glucose to fuel the brain and build glycogen reserves. In addition, the influx of carbohydrates and protein provides extra intermediates for the CAC, and the protein also contributes to muscle building.

carbohydrates (derived from the meal's carbohydrates and protein) can also be stored as glycogen. Plus, the energy and protein from the meal can contribute to the building of muscle tissue. Not surprisingly, since we've eaten a big meal, our hunger goes away, too.

We'll enjoy a number of hours in the fed state, but eventually the energy from the big meal will run out. When it does, blood sugar starts to drop. Amino acids derived from muscle start to build up, and we begin to feel hungry. The body begins breaking down glycogen to help maintain the brain's supply of glucose. Glucose from the breakdown of glycogen, and fatty acids from the breakdown of stored fat are converted into acetyl-CoA to fuel the CAC. Glucose from glycogen and amino acids is also converted into oxaloacetate. Amino acids also feed into oxaloacetate directly as well as into other CAC intermediates.

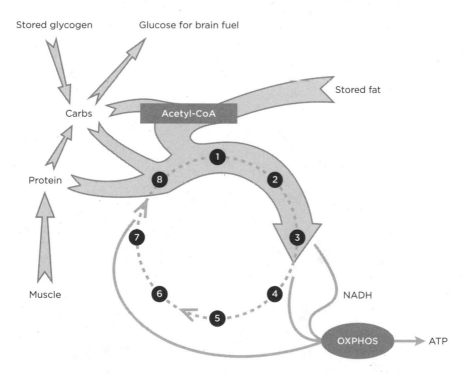

Figure 9.4. When the energy from a meal runs out, the body starts to draw on stored glycogen and muscle to provide glucose (from carbohydrates). Stored fat can also be converted to acetyl-CoA. Energy flow through CAC decreases.

We can cancel that hunger with shivercise or tightercise, which creates a huge demand for energy that signals the body to secrete epinephrine, glucagon, growth hormone, and cortisol in order to quickly mobilize both glycogen and fat stores to increase ATP production. As shown in figure 9.5, this allows more conversion of glucose to acetyl-CoA and oxaloacetate. More fatty acids from stored fat are also released and converted to acetyl-CoA. Interestingly, muscle contractions also release calcium, and calcium stimulates the third and fourth reactions of the CAC. Stimulating the fourth reaction (which you'll recall was the bottleneck when NADH is high) helps the CAC speed up. At the same time, calcium stimulates the entire OXPHOS cascade, increasing ATP production to help meet the energy needs of exercising muscle.[26] Since plenty of glucose is released from stored glycogen to supply the brain with fuel, muscle breakdown stops, and hunger pangs go away.

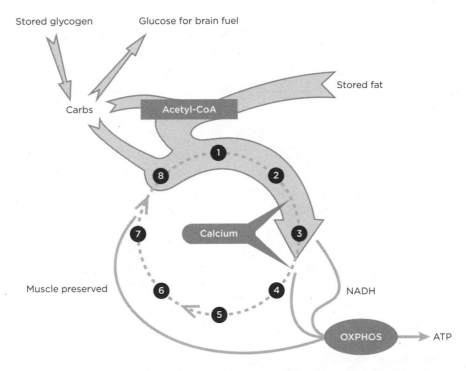

Figure 9.5. Fastercise mobilizes stored glycogen and fat to provide sufficient fuel for the CAC without having to degrade muscle. Muscle contractions result in a release of calcium, which stimulates some CAC reactions.

To discover for yourself how quickly these changes can happen, the next time you start feeling hungry, try tightening as many of your muscles as you can, as hard as you can, for a few seconds. By the time you let your muscles relax, or soon thereafter, you will probably notice that the sensation of hunger has decreased. As discussed earlier in the book, yawning (or taking a deep breath) may have a positive effect, too, because it expels CO_2 and provides a little more oxygen for ATP production.

Each time you shivercise or tightercise, you use up some of your body's glycogen stores to burn some fat, fuel your brain, preserve your muscle, and cancel your hunger. Your hunger might stay away for one or two hours, but when your glycogen becomes sufficiently depleted, your hunger will tend to return within fifteen minutes. When fastercise stops keeping your hunger away, you can eat another high-carb meal (remembering to fastercise just before you start eating, too) in order to replenish your glycogen stores. And while your body is in the fed state, it can also use the protein supplied by the meal to build muscle.

High-Carb Approach: The Fat Pump

I liken the integration of fastercise and the high-carb approach to a pump. The body is pumping off fat intermittently, as one would pump water with an old-fashioned hand pump. When we operate a hand pump, water is expelled only when we push the handle down, not when we pull the handle up. But both the down and the up strokes are needed to continue to pump the water. Similarly, with the high-carb approach, the body burns fat when glycogen levels go down but not when glycogen levels are pushed up. Glycogen levels in muscle tissues and the liver increase (get replenished) during the fed state, which is not conducive to fat burning. Then we fastercise when hungry (in the non-fed state) and our glycogen goes down, releasing just the right amount of glucose to help us burn fat and preserve muscle. Likewise, in the fed state insulin helps us build our glycogen and muscle and our bodies become more sensitive to the counterregulatory hormones. Then, in the non-fed state the counterregulatory hormones help us burn fat while our insulin sensitivity increases. Thus, for every cycle of our fat pump our glycogen level is pushed up and pulled down once, and some of our stored fat is "pumped" out.

A person who has adapted to a high-carb diet can burn carbohydrates more easily and fat less easily than a person who is low-carb-adapted.

Consequently, that person tends to burn through glycogen faster and can't easily access stored fat as a backup energy source. As a result, a high-carb-adapted person tends to get hungry more abruptly when the fed state comes to an end. On the other hand, a low-carb-adapted individual tends to burn through glycogen more slowly and the onset of sensations tend to be less abrupt because that person can more easily access stored fat for energy. Glycogen is an easily exhausted supply of stored energy; on the other hand, the only way we'll run out of stored fat as an energy supply is by losing all our fat, which might take weeks of fasting.

Following a high-carb diet may make it a little harder to burn up fat stores. We've seen in the research above that people can lose fat on a high-carb diet, but the fat loss tends to happen at a slower pace than for people on a low-carb diet.

In my opinion, rapid onset of hunger indicates a rapid breakdown of muscle. When the body is high-carb-adapted, it does not have high ketone levels. When carbohydrates from glycogen are depleted and there are insufficient ketones to support brain function, the body has no choice but to turn to muscle in order to make carbohydrates to supply glucose for the brain. Low ketone levels is why our muscles can be more vulnerable to muscle loss when we are on a high-carb diet versus a low-carb diet, especially in circumstances where we don't have ready access to food (to replenish our glycogen) and we can no longer push off hunger with fastercise due to depleted glycogen levels.

Low-Carb Approach: The Fat Siphon

I liken the integration of fastercise and the low-carb approach to a siphon because the body can burn fat and build muscle more continuously than with the high-carb approach. A siphon is a tube that's used to draw a liquid, such as water, from a higher reservoir down to a lower level. A siphon has to be set up correctly or it won't work well. The amount of flow depends on the diameter of the tube. The outflow end of the tube has to be positioned lower than the inflow end. To induce flow through the siphon tube, the inflow end of the tube needs to be filled with water, usually by suction or immersion, at least to the point that the column of water descending the tube extends lower than the surface of the higher reservoir. Sometimes, the siphon needs to be adjusted to maintain the flow. For example, as the water

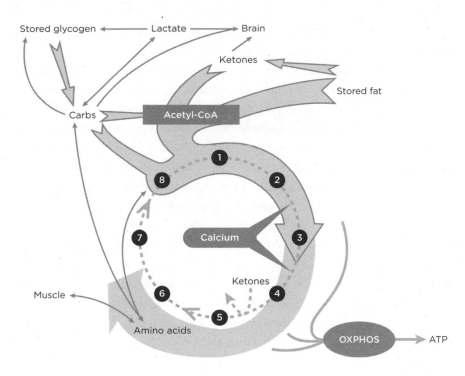

Figure 9.6. When the body has adapted to a low-carb diet and a steady supply of fat is powering the CAC, there is enough extra capacity not only to power ATP production but also supply energy for the building of proteins.

level in the reservoir drops, the inflow end may have to be repositioned to make sure it remains underwater and doesn't start taking in air, which can cause the flow of water to stop.

The siphon analogy fits the low-carb approach to eating in a surprising number of ways. For example, lowering carbohydrate intake draws more fat into the siphon. Fat stores more than twice as much energy as muscle. Energy flows through the "tube" from the higher energy content of fat to the lower energy content of muscle. The tube in this case is the CAC and OXPHOS.

Reducing carbohydrate intake to increase ketone levels is another aspect of setting up the siphon. Ketones help the fuel continue to flow through the tube by sparing carbohydrates to supply a necessary intermediate in the CAC (so the siphon doesn't start "sucking air"). Ketones also help the CAC turn. It can take two to seven days for blood ketones to rise above 0.5 mM, which is the level at which the body officially enters ketosis. The fat siphon continues to develop over two to six weeks as our machinery and pathways adapt to rely more on fat for fuel.

Fastercising amplifies the siphon effect, too. Fastercising when hungry draws upon the body's glycogen and stored fat as fuel to supply the needed ATP. This fills the siphon tube with carbohydrates from glycogen and fatty acids from fat and initiates a strong flow. Fastercise also releases calcium and stimulates ketone production, both of which help the CAC turn. Calcium also helps energy flow through OXPHOS to make ATP. And as we know, fastercise also increases the size and number of mitochondria in the body, which is like increasing the diameter of the siphon tube. By mobilizing glycogen stores, fastercise also adds lanes of traffic to the CAC roundabout, which can also be likened to enlarging the siphon tube.

Although the body isn't breaking down fat and converting it directly into muscle, the energy from the breakdown of stored fat can be used to help build muscle. My own experience satisfies me that my body can lose fat and build muscle on a meal-by-meal basis with either the high-carb or low-carb approach. However, since the low-carb approach has a more continuous feel to it than the high-carb approach, I liken it to an effectual siphon that can convert fat, in a sense though not directly, to muscle.

We can also compare the fat siphon generated by fastercise and a low-carb diet to the way we access energy during a long-distance running event. Depending on the distance, runners usually don't eat during a race, but their muscles are contracting throughout, and they are using up energy, yet they don't feel very hungry. In a similar way, when we fastercise to extend the time between meals, we are going for fairly long periods of time without eating. We contract our muscles in small doses from time to time, just enough to keep the siphon running and tamp down hunger. It's as though big parts of our day can become like a drawn-out sporting event.

To summarize, the low-carb approach affects the body more like a fat siphon and less like a fat pump because it spares glycogen. Since glycogen stores don't run out as abruptly with the low-carb approach, there is less need to interrupt the burning of fat in order to refill glycogen stores with carbs. The low-carb approach spares glycogen in many ways, including:

- More acetyl-CoA is supplied by fatty acids (from food and storage) rather than carbohydrates.

Lactate's Effect on the Fat Siphon

Lactate stimulates an increase in the production of monocar-boxylate transporters, which are molecules that can transport both lactate and ketones across cell membranes and mito-chondrial membranes.[27] These transporters are not hormonally regulated, which makes it possible for lactate and ketones to enter into cells and mitochondria very quickly without the need for insulin. This process enables the fat-burning siphon to keep running smoothly even when insulin levels are low, which is significant because it enables the body to keep burning fat to make ATP. That ATP can power the recycling of lactate into glucose, glucose into glycogen, and amino acids into muscle even when the body is not in the fed state.[28] The party doesn't have to stop just because insulin levels go down.

- More acetyl-CoA is supplied by ketones rather than carbohydrates.
- Ketones replace some glucose as fuel, promote the recycling of carbohydrates.
- Dietary amino acids replace some glucose as fuel, provide anaplerosis for the CAC.
- Energy from fat is used to recycle carbohydrates back into glycogen.

The Importance of Water

Water greatly facilitates the transport and burning of fat.[29] Think about it: How do we transport fat off our dirty dishes after dinner? We use soap . . . and water. If we tried to use soap without water to clean our dishes, we'd just make a greasy mess. But with soap *and* water we can transport the fat off our dishes quite easily until they are squeaky clean. It's the same way in our bodies.

Intramuscular Triglyceride (IMTG)

The body uses intramuscular triglyceride (IMTG) in the low-carb approach in a similar way that it uses muscle glycogen in the high-carb approach. Muscle glycogen is a form of carbohydrate that is stored within the muscle itself (as opposed to the liver) for easy access. IMTG is stored as droplets of triglycerides in muscle cells (as opposed to adipose tissue) for easy access. However, as we follow the high-carb approach, over time we increase our dependence on muscle glycogen for energy, and as we follow the low-carb approach, we increase our dependence on IMTG for energy.[30]

People who regularly do endurance exercise burn more fat than those who don't.[31] This makes sense because endurance exercise depletes glycogen so that the exerciser must rely more on fat for energy. Endurance exercise is like an accelerated form of fasting because they both involve depleted glycogen stores and increased dependence on fat for fuel. Fastercise with the low-carb approach also involves depleted glycogen stores and can train our bodies to rely a little more on IMTG and a little less on muscle glycogen. Well-trained athletes can store as much energy as IMTG as they do as muscle glycogen.[32] Ketones can increase the burning of IMTG even when glucose and insulin levels are high.[33] This is strong evidence that ketones do indeed help us burn fat. However, when we ingest extra ketones, or make more from a source such as medium-chain triglyceride oil (MCT oil), there is a chance that we will burn a little less stored fat temporarily because the extra ketones will be burned for fuel more easily than fat.

We have all heard about lipoproteins such as HDL and LDL and how important they are in assessing our risk for heart disease. These lipoproteins work like soap in the body. Portions of them mix well with water, and portions of them mix well with fat. The body employs these lipoproteins to transport cholesterol and fat through water (blood) to our cells so the fat and cholesterol can then be transported across the cell membranes and the mitochondrial membranes and burned to release energy (or used for other purposes). Incidentally, all of our cell membranes and mitochondrial membranes are made up largely of soap as well. They are made of phospholipids that have a portion that mixes well with water and a portion that mixes well with fat. All of this beautiful soap works much better when we have plenty of water to go along with it, just as with our dishes. From that context, it's easy to see why hydration is a very big deal when it comes to fat loss. This may also explain the metabolically clean feeling I get sometimes when I fastercise and hydrate when hungry.

Dehydration is also antiketogenic![34] This is a point about water in ketogenic diets that we don't hear mentioned very often. We hear about how we can more easily become dehydrated on a ketogenic diet but not that dehydration lowers ketones and lowers the number of calories coming from fat. On a ketogenic diet, hydration increases the number of calories coming from fat and increases ketones in the blood. For convenience, some people like to mix electrolytes into their water and drink that throughout the day.

Here's one example from my own experience. Once, I had a blood ketone level of about 0.3 mM and I tightercised and I got hungry again in ten minutes. Then, I tightercised again and drank a lot of water and wasn't hungry again for ninety minutes. At the end of those ninety minutes, my ketone level was 1.2 mM. Other times my ketone level has gone up about 0.6 mM in thirty minutes after I tightercised, took supplemental amino acids, and drank water. Tightercise alone, though, tends to lower ketones for a time because ketones can be used as fuel.

Water not only helps keep the siphon running; it also helps decrease our hunger because of the stretch receptors in our stomachs that will decrease our hunger as the stomach fills up with water. The stretch receptors detect distension of the stomach and send signals directly to the brain to trigger a reduction in appetite. However, this effect disappears as soon as the distension is gone.[35] In my own example above, my appetite remained at bay for ninety minutes and I'm sure my stomach was drained of water much sooner than that. One way to gauge how much water to drink is to treat your hunger as if it were thirst. When you get hungry, try doing shivercise or tightercise and then fill up on water. For example, once I was quite hungry and I drank about thirty-two ounces of water (without fastercising), and two minutes later my hunger went away. About thirty-five minutes later, it swiftly returned.

Effects on Fatty Acids

Ketone levels rise with exercise as fatty acid levels rise, and ketones seem to increase more with high-intensity exercise in people who are less fit.[36] At the same time, ketone use by muscles increases fivefold with exercise.[37] Whether ketone levels go up or down with exercise depends on whether or not ketones are being used up faster than they are being produced. Intense exercise can use up ketones quickly, lowering them at first, followed by a rise.[38] It's as though using up the ketones quickly gets the body's attention (sends a signal), which leads to release of fatty acids. Ketone production then rises. Exercise has a greater effect on fatty acid and ketone production in people who are low-carb-adapted.[39] Thus, the more low-carb-adapted we become, the better we get at burning fat and making ketones.

As fatty acids rise, we make more uncoupling proteins. Uncoupling proteins allow protons that have been pumped into the intramembrane space of mitochondria to flow back into the mitochondrial matrix without making ATP. This process releases a lot of energy as heat, which helps explain why we can feel more heat released when we fastercise on the low-carb approach versus the high-carb approach. As crazy as it is, many people find it easier to lose fat by canceling hunger with shivercise on a low-carb diet than they do running for miles every day. As it turns out, running long distances can actually slow the metabolism, especially for obese people. Running can use up our glycogen, which can make our protein more vulnerable. We can work up an appetite, crave carbohydrates, overeat, and not lose fat, and might even gain fat. On the other hand, we can get the fat-burning siphon started by getting hungry and fastercising, which can fill the CAC with intermediates and release calcium to help the CAC turn. To keep the siphon turning we can implement a low-carb diet with meal swinging by contracting our muscles when we start getting hungry, fastercising right before we eat, and eating lightly when we do eat. Stretching (contracting) hard for a few seconds at the right time under the right circumstances might accomplish more than running for miles. It's counterintuitive.

Summing Up the Low-Carb Approach

With the low-carb approach, we can design our meals with the goal of feeding the siphon rather than interrupting it. If you're already in ketosis, focus on providing some amino acids by eating some protein (meat, fish, eggs, dairy, or vegan). Also, add water and electrolytes (green vegetables). The vegetables will supply plenty of carbohydrates, and you can eat some fat as well to satisfy your appetite as you wish. If you eat lightly at meals you will get hungry again sooner and can answer your hunger between meals with tightercise and water. The point here is that even meals can feed into the fat siphon without disrupting the fat siphon very much.

With the low-carb approach, you don't need to eat much carbohydrate, because your body can rebuild your glycogen stores (your body still uses glycogen even when you are low-carb-adapted) by recycling glucose even when your insulin levels are low. And you can do well on eating about half as much protein as on a high-carb diet because your body develops a greater capacity to recycle amino acids. You don't even need to eat a lot of fat, because you'll be burning your stored fat. In fact, you could actually stop eating completely for days if you wanted and the siphon would still run. That would be called an extended fast. You would need to drink water regularly, but your body is capable of surviving without food for many weeks. The point here is that the fat siphon is already built into our bodies, and we can understand it and work with it to easily siphon stored fat out of our bodies.

A low-carb diet increases the level of a branched-chain amino acid (BCAA) called leucine, which stimulates protein synthesis even when insulin is low. Thus, we don't need to interrupt the fat siphon to increase insulin in order to continue building muscle. We can keep building muscle even when the fat siphon is running and insulin levels are low. The levels of leucine and other BCAAs rise because they are spared as the body burns ketones instead.[40] Rising levels of BCAAs also helps explain why there is less need to eat protein when on a low-carb diet; in fact, eating too much protein can be antiketogenic.

Exercise and amino acids both promote muscle growth, and ketones preserve and encourage muscle growth, too. When you are in ketosis your body recycles a lot of protein. If you add just a little protein through your diet,

Low-Carb Approach Burns Fat, Replenishes Glycogen

One recent study showed that low-carb-adapted endurance athletes burn 2.3 times more fat than endurance athletes who are high-carb-adapted, even though their glycogen levels before exercise, after exercise, and after recovery showed no significant differences.[41] This demonstrates that once we are low-carb-adapted our bodies burn fat more easily and are able to replenish glycogen stores equally as well as they would while on a high-carb diet.

it can contribute significantly to muscle synthesis. Therefore, fastercise, amino acids, electrolytes, water, and a state of ketosis can be a great combination.

Fat burning is optimized when just the right amount of carbohydrate is present. If the carbohydrate availability is too low or too high, fat burning can be inhibited. Trying to hit that sweet spot manually with food would be impossible, but fortunately, we can replenish our glycogen and our bodies can figure out the calculations and handle it all automatically. Our bodies are built to deliver us the maximum amount of energy possible in order to help us survive by foraging. When we answer hunger with fastercise, our bodies burn just the right amount of glycogen and just the right amount of fat to provide us with maximum energy. Perfect. When we answer hunger with fastercise once or twice in the morning before eating it may seem as though we are skipping breakfast.[42] But in a sense, we have eaten breakfast; it's just that we ate it the day before. The energy stored in our glycogen becomes our "breakfast."

Diets, Meal Swinging, and Fastercise

As you recall, meal swinging involves pushing off hunger one or more times with fastercise before we eat again. But what is the best combination of high-carb, low-carb, shivercise, and tightercise if our main goal is to lose fat, or to build muscle, or a balance of the two? The research hasn't been done yet to answer that question definitively, but here I offer some ideas to consider in the meantime, and my own current thinking.

The research reviewed above suggests that low-carb diets can burn more fat while high-carb diets can build more muscle, and that low-carb diets provide the biggest changes in body composition. That is in line with my current understanding and personal experience. From a fastercise stand-point, if our main goal is to lose fat, it stands to reason that the more we push off hunger with fastercise, the more time we spend in the non-fed state and the more fat we will lose. However, I encourage you to pay close attention to your appetite and never let it go away on its own. Either push it off with fastercise or fastercise and eat lightly. Dwelling in the less-fed state (going around hungry) will likely dampen your meal-swing momen-tum, or in other words, slow your metabolism.

I believe that using the non-fed state and fed state to enhance each other by increasing our sensitivity to the hormones involved can produce signifi-cant changes in our bodies. Therefore, I like the idea of pushing the swings as high as possible. I tend to cancel hunger with fastercise as many times as I can until it's no longer sufficiently convenient or effective. In other words, I like to fastercise until my hunger is pretty insistent. Then I reach for the cramp with tightercise and then I eat lightly and stop when I'm just full.

I use both shivercise and tightercise to push off hunger. I do recommend shivercising at least once a day until you catch a deep breath in order to stimulate the growth of your mitochondria every day. Beyond that, shiver-cise and tightercise both have their advantages and disadvantages for the purpose of pushing off hunger.

I tend to use shivercise if I need to increase my mental focus, energy, or want to have a fun break listening to some energizing music and livening up the day. In just a minute or two shivercise can really help you clear your mind and charge up your day. Shivercise tends to use up ATP and glycogen faster than tightercise. Shivercise may push off your hunger more effectively, especially when you are high-carb-adapted, but the sooner your glycogen is depleted, the sooner fastercise will stop being very effective for pushing off hunger. Also, shivercise uses up glycogen so fast that it might interfere with your fat siphon if you are using the low-carb approach.

What's Your Why?

Why do you want to get in better shape and live a healthier life? Identify that reason and remember it. I believe this will be crucial to your success because even though your body will respond automatically to the signals you send, you still need to consciously send them. If we lived in a hunter-gatherer society, then perhaps you would need to hunt and forage before you ate, and perhaps you would naturally be eating real unprocessed foods. But in a society with easy access to lots of processed convenience foods, sending your body the right signals and making the right eating decisions takes conscious thought and resolve. As easy as fastercise signals are to send, consciousness requires motivation, which requires a why. Consider writing down your why and posting it in one or more places where you will see it every day. Track your observations, signals, and progress in a journal. Doing so will enhance your success.

On the other hand, tightercise works quite well to gently push off hunger when you are using the low-carb approach. Tightercise may be more effective at increasing muscle size and strength. It can be done extremely conveniently whenever you notice the slightest twinge of hunger. It fits easily into your normal daily activities. You don't need to go to another room or even get out of your chair, and it can be done inconspicuously in public settings. Your body sends you the hunger signal and you reply by tightening your muscles hard for a few seconds to direct your body to snack on storage. Easy peasy. Since tightercise spares glycogen more than shivercise does, with tightercise you can usually go longer before having to eat again. I believe that as you continue to build up your mitochondria with regular shivercise and tightercise, over time you will find tightercise increasingly effective at pushing off your hunger.

So what is the best combination of high-carb, low-carb, shivercise, and tightercise? Maybe the best answer is the combination that fits best into your lifestyle, the one that you enjoy doing and can easily continue. Another good answer is the combination that shifts your survival balance furthest into forage mode, which is characterized by more muscle and less fat. Thus, it may be that the signals that build your muscle the most will be the same ones that burn your fat the most, and vice versa. That's why I feel that sending the forage signals with meal swinging is the best way to shift your body into forage mode and to shift your fat set point:

1. Cancel hunger with fastercise.
2. Fastercise immediately before you eat. (I usually shivercise or tightercise the muscles I want to build.)
3. Eat lightly (just until full).

Simple.

CHAPTER 10

The Promise of Fastercise

C oming full circle, in this final chapter of the book I return to the topic of Wilson's temperature syndrome (WTS), which can leave people suffering with symptoms of slow metabolism even though their thyroid blood test results are in normal range. As I described in the introduction, WTS tends to come on or worsen at times of emotional, mental, or physical stress such as childbirth (the most common cause), divorce, the death of a loved one, and job or family stress. Over the past twenty years, I've trained thousands of physicians on how to correct WTS by normalizing low body temperature patterns using a T3 therapy protocol that I developed. This successful treatment approach has stood the test of time, and doctors all over the world are using this therapy to help their patients recover.

Although WTS can be quite debilitating, people can often recover from it completely such that their temperatures and symptoms remain improved and they do not have to take medication indefinitely. It may be that this recovery is possible because WTS is simply a manifestation of the survival balance being set too far into storage mode. Why is WTS so prevalent now? It may also be a manifestation of the explosion in modern conveniences and the resulting drop in foraging behavior. I am excited about the potential for fastercise helping many people who suffer from WTS to normalize their temperatures and recover by shifting the body's survival balance more toward forage mode. To show how this may be possible, I'll first briefly explain the similarities and differences between hypothyroidism and WTS.

WTS is a syndrome, not necessarily a disease. In a sense, the body goes into conservation mode to help deal with stress or starvation. Sometimes, the body shifts into this mode so far and for so long that it becomes the new normal, maladaptive and abnormally persistent. It can shift the body's survival balance set point too much toward storage mode. Low body temperature and slow metabolism can help the body avoid starvation, but they also can make normal functions of daily life more difficult. And unless a person with WTS can signal their body to shift the survival balance more toward forage mode, that person can remain stuck in conservation or storage mode. In fact, when thyroid blood tests are normal, average body temperature may be the most revealing indicator of the survival balance set point.

The thyroid system plays the central role in helping the body maintain a normal metabolic rate, or in other words, a normal body temperature. Both hypothyroidism and WTS cause slow metabolism and lower body temperature. In hypothyroidism, the thyroid gland does not produce an adequate supply of thyroid hormones, and this can be confirmed by thyroid blood tests. This condition is often permanent, and people with this condition typically need to take thyroid medicine for life. On the other hand, people with WTS can have the same symptoms of slow metabolism and low body temperature even though their thyroid hormone supply and thyroid blood test results are normal.

Though the incidence of hypothyroidism is on the rise in the United States, I estimate that WTS is about ten times more common. Frank hypothyroidism is estimated to affect about 1 percent of the US population, with some authors estimating it affects almost 4 percent.[1] By contrast, I estimate that WTS affects 30–40 percent of the population. I've found that herbal and nutritional supplements and T3 therapy support people in their recovery from WTS as they shift their survival balances away from storage mode. Fastercise may be another technique that helps people accomplish this change.

The Temperature–Metabolism Relationship

What are the relationships between thyroid function, temperature, survival balance set point, herbs, nutrients, T3 therapy, and fastercise? The answer to this question is multifaceted. For our bodies to function optimally, it is important that their metabolic rates be neither too high nor too low. Metabolic rate is determined by the pituitary and thyroid glands, as well as by the tissues of the body.

Symptoms of Slow Metabolism
(Low Body Temperature)

The following list includes many, although not all, of the symptoms that people who have WTS or hypothyroidism may experience:

Fatigue
Headaches
Migraines
Easy weight gain
Premenstrual syndrome
Irritability
Fluid retention
Anxiety
Panic attacks
Hair loss
Depression
Faulty memory
Difficulty concentrating
Heat intolerance
Cold intolerance
Abnormal swallowing
  sensations
Insomnia
Constipation
Low motivation
Lack of ambition
Irritable bowel syndrome
Muscle aches

Aching joints
Dry skin
Dry hair
Hives
Asthma
Allergies
Brittle nails
Slow healing
Sweating abnormalities
Carpal tunnel syndrome
Raynaud's phenomenon
Itchiness
Irregular menstruation
Acne
Low libido
Easy bruising
Unhealthy nails
Ringing in the ears
Flushing
Bad breath
Dry eyes
Blurred vision

The thyroid gland produces a hormone called thyroxine (T4), which is converted to triiodothyronine (T3). T3 is the active form. Some T3 is produced in the thyroid gland but 80 percent of it is produced inside the cells of the tissues of the body. Thyroid blood tests can measure how much T4 the thyroid gland supplies to the blood, but the tests can't measure how well that T4 is being converted to T3 inside cells. T3 enters the nucleus of the cell and tells the cell how fast to transcribe DNA. Since DNA is the code of life, T3 is a key factor in determining how fast we live. The faster our bodies transcribe DNA, the faster proteins are synthesized and the faster chemical reactions take place (that is, our metabolic rate increases). Thus, it's no surprise that the body doesn't function well when the effect of T3 is insufficient, whether that is due to insufficient production of T4, poor thyroid hormone transport into the cells, poor conversion of T4 to T3, poor transport of T3 into the nucleus, poor engagement of the T3 with the thyroid hormone receptor, or some other problem.

Maintaining body temperature within a narrow range is important because the progress of chemical reactions is affected by temperature, primarily because temperature affects the action of enzymes. Without proper enzyme function, chemical reactions can't proceed optimally. Temperature affects the physical shape of molecules, and molecular shape affects how well enzymes can function. When enzymes are too cold, their shape becomes too tight, and when they are too hot, their shape becomes too loose. When temperature is just right, enzymes function optimally. The shape of enzymes is also determined by their structure, which is determined by our genetic code. It's significant that the body characteristic that is most similar among all humans is temperature. Humans function best at similar body temperatures because the structure of their enzymes are so similar. Whether young or old, obese or lean, clothed or naked, sleeping or exercising, human bodies maintain remarkably similar temperatures. This fact is why humans everywhere can use a standard thermometer to measure their body temperature, and why so many people feel best when their body temperature is 98.6°F (37.0°C). Body temperature is much less variable from one person to another than height, weight, hair color, cholesterol level, blood pressure, or any other parameters one might suggest.

Which comes first? Does lower body temperature lead to slower chemical reactions or do slower chemical reactions lead to lower body temperature? The answer is—both. Body temperature is a *measure* of metabolic rate. A thermometer is like a speedometer. The thermometer reading increases as the speed of the molecules it is measuring increases. When a thermometer shows that the temperature outdoors is getting warmer, it means that the air molecules surrounding the thermometer are speeding up. Think about what happens when you put a pot of water on the stove and turn on the burner to boil the water. The burner starts generating heat and then the water starts to heat up. The molecules of water in the pot move faster and faster. Put a thermometer in the pot, and you'll see that as the temperature continues to rise, some of the water turns into water vapor, and bubbles start forming. You can watch as the formation of water vapor and bubbles speed up, boiling begins, and then it speeds up to a rapid boil. Similarly, your body temperature indicates how fast your body is operating. The metabolic rate is directly proportional to temperature for all forms of life on earth.[2] This is true for cold-blooded animals, warm-blooded animals, plants, and microorganisms. When temperature goes up, metabolic rate goes up. When temperature goes down, metabolic rate goes down, independent of any form of regulation or any other factors.

Researchers at Williams College demonstrated this in a study of the metabolism and temperature of a lab mouse. We know that oxygen consumption is a direct measure of metabolic rate because the faster we live, the faster we consume oxygen. Body temperature is also a direct measure of metabolic rate because the faster we live, the more heat we generate. The researchers inserted a temperature sensor into the abdomen of a mouse. This sensor transmitted its temperature readings wirelessly to a receiver for data collection. At the same time, the researchers measured the amount of oxygen the mouse consumed from its enclosure over time, as shown in figure 10.1. Notice how closely the mouse's body temperature mirrored its oxygen consumption. The mouse's body temperature rose almost immediately after oxygen consumption increased.

You may have wondered whether you have a slow metabolism. You may have even consulted a doctor to find out. Many doctors respond to questions and concerns about metabolic rate by ordering thyroid blood tests. However, such blood tests do not measure metabolic rate, because they do not measure oxygen consumption or body temperature. They measure only the levels of

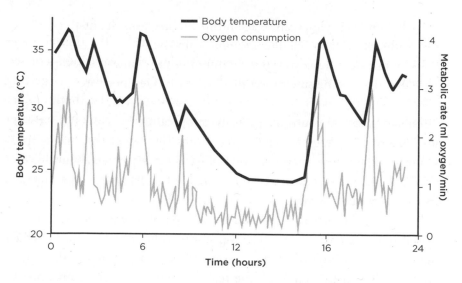

Figure 10.1. The changes in core body temperature of a mouse correlate closely with the mouse's oxygen consumption, and thus, with the mouse's metabolic rate. Adapted from S. J. Swoap, "The Pharmacology and Molecular Mechanisms Underlying Temperature Regulation and Torpor," *Biochemical Pharmacology* 76, no. 7 (2008): 817–24.

thyroid hormone in the blood. It's true that thyroid function is critical for normal metabolism, and low thyroid function can certainly be the cause of low body temperature or slow metabolic rate. But a normal result from a test such as the thyroid-stimulating hormone test by no means guarantees a normal metabolic rate. In fact, by far, most people who experience low body temperatures and corresponding symptoms have normal thyroid-stimulating hormone levels and are suffering from WTS, not hypothyroidism.

Measuring Body Temperature

Fortunately, you can measure your own metabolic rate more accurately with an ordinary thermometer than doctors can with thyroid blood tests. I recommend the following protocol:

1. Using a glass liquid-metal thermometer (such as Geratherm), take your temperature three hours after you wake up and record the reading.

2. Measure your temperature again three hours after the first reading, and then again three hours after that.
3. Continue checking your temperature in this manner for several days in a row. Be sure to record all times and temperatures.

You can calculate your average daily temperature by adding the three individual readings together and then dividing by three. I find that people usually feel their best when their oral temperatures average about 98.6°F (37.0°C). You can download a temperature log for charting your temperatures at www.wilsonssyndrome.com/handy-tools/.

Figure 10.2 shows the results of a study comparing the temperature readings of nine obese volunteers and twelve lean volunteers. Notice that the temperatures of all volunteers were essentially the same through the night and first thing in the morning. The real difference in their temperatures started several hours after they woke up and lasted for about six hours. This corresponds to my experience with patients, and that's why I recommend the specific protocol for measuring temperature outlined above.

The results of this study suggest that if people can shift their survival balances toward forage mode and greater leanness, their daily temperatures may also rise. Knowing the relationship between metabolic rate and body temperature, I was curious to see how fastercise affects body temperature, so I used myself as a test subject. I measured these temperatures by swallowing an electronic pill that wirelessly transmitted my body temperature readings every six seconds to a data recorder.

Figure 10.3 shows the changes in my body temperature. Around 7 a.m. I took a bike ride for about an hour and a half. Notice how my temperature went up during the bike ride and came back down again when I stopped. At around 10:35 a.m. I started getting hungry, so I did one minute of shivercise. My hunger went away and my temperature started rising. About half an hour later I started getting hungry again, so I shivercised again for a minute. Again, my hunger went away and my temperature continued rising. Then, half an hour later my hunger started coming back and I shivercised again for a minute. Again, my hunger went away and my temperature continued rising. What's interesting is that although my temperature dropped right away after biking, my temperature didn't drop back down again as soon as I stopped fastercising. Those three minutes of shivercise affected my metabolic rate more

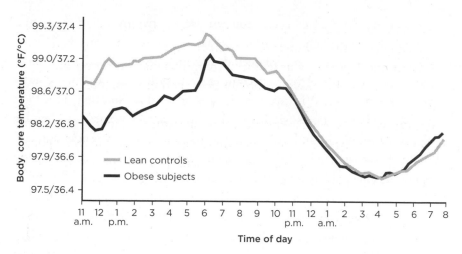

Figure 10.2. A study of obese and lean volunteers shows that obese subjects tended to run lower body temperatures during the day compared to lean controls. Adapted from Grimaldi et al., "Evidence of a Diurnal Thermogenic Handicap in Obesity," *Chronobiology International* 32, no. 2 (2014): 299–302.

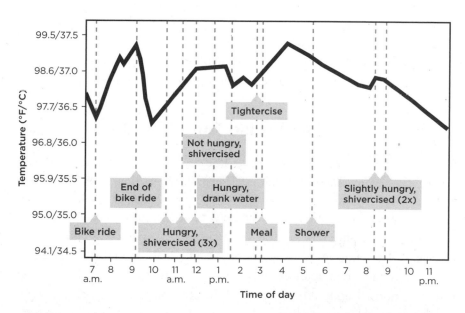

Figure 10.3. The results of twenty-four hours of continuous monitoring of the author's body temperature.

than ninety minutes of biking did. You can also see that my body temperature increased by eating lunch. (I didn't eat breakfast or dinner on this day.)

Eating can be very mood enhancing. When you're famished and you eat some food, you can feel yourself coming back to life as your total energy expenditure increases. When you eat, heat is released as your body uses food to *build* your energy stores (similar to the way a battery warms up when it's charging) and to make and burn ATP for immediate energy production and use. On the other hand, when you fastercise, heat is released as your body *mobilizes* its energy stores and makes and burns ATP to power itself (similar to the way a battery heats up as it discharges). In both situations, your appetite goes away because your body is flush with fuel, in one situation from food, in the other from storage. This demonstrates how a bout of fastercise can replace a bout of eating to provide energy, eliminate hunger, and increase metabolic rate. When we don't have access to fuel either from food or fastercise, our temperatures tend to go down and we can get hungry, cold, tired, weak, dizzy, and sleepy, and we may begin daydreaming about food. Not surprisingly, these symptoms of low temperature can also be brought on by low thyroid function.

Eating promotes the building of fuel stores in the body, conserving energy for later. As explained earlier in this book, signaling the body to conserve energy more and to forage less shifts the survival balance more toward storage mode. By helping the body shift the survival balance more toward forage mode, fastercise holds the promise of helping people escape storage mode and change their survival balance set points, or their fat set points.

Fastercise for WTS

Because WTS is so prevalent in our population, it stands to reason that many of you reading this book also have low body temperatures and would show "normal" results on thyroid blood tests—in other words, many of you may be experiencing WTS. Fortunately, there are lifestyle measures that can help WTS sufferers support normal body temperatures and recover. Since WTS tends to be brought on by physical, mental, or emotional stress, anything that reduces stress and promotes good health can help support normal body temperatures. A healthy diet, regular exercise (you can't get much more regular than fastercising when hungry and before every meal), and adequate sleep are all excellent measures to take. Ensuring adequate intake of certain key nutrients is also important. The enzyme that converts T4 to T3 and helps combat autoimmune disease of the thyroid gland contains selenium. The *3* in T3 stands for iodine. There are three iodine atoms in every molecule of T3.

Including certain herbs in your diet can help supply these nutrients. For example, kelp contains iodine, which is also an important nutrient for other tissues such as the breast. The roots of blue flag (*Iris versicolor, I. virginica*) have been used for centuries to promote healthy drainage of the thyroid gland. (We can think of this in terms of detoxification.) Guggul (*Commiphora myrrha*) may enhance the uptake of iodine, the production of thyroid hormone, and the conversion of T4 to T3.[3] Adrenal support herbs such as *Eleutherococcus senticosus* and ashwagandha (*Withania somnifera*) help the body deal with stress, and thereby help support thyroid function and normal body temperatures. Many of the herbs that help support healthy thyroid function are anti-inflammatory. This makes sense, since the thyroid gland is the organ in our bodies that is most susceptible to inflammation and is the organ most often attacked by autoimmune disease.

Although many people can recover from WTS using diet and lifestyle measures alone, others may need T3 therapy. T3 is a thyroid hormone that must be prescribed by licensed practitioners. T3 therapy involves replacing some or all of the T4 in a person's body with T3 for a time. You can think of it as a way to reboot the thyroid system by clearing out the thyroid pathways and letting them come back online with a fresh start. T3 therapy is designed to normalize low body temperatures and to reset the thyroid system so that the body can maintain normal temperatures on its own even after the T3 therapy has been discontinued. T3 therapy can be used in patients who have normal thyroid gland function, or in those who have hypothyroidism. T3 therapy involves cycling on and off T3 therapy one or more times.*

Part of the promise fastercise holds in helping to correct WTS is the connection between thyroid function, epinephrine, mitochondria, and

---

* To find a practitioner certified in the Wilson's T3 Protocol that I developed, check online at www.wilsonssyndrome.com/patients/medical-providers/.

fastercise. As discussed in chapter 7, thyroid hormone stimulates some of the same cascades stimulated by epinephrine and increases the density of epinephrine receptors, and epinephrine promotes T4-to-T3 conversion. The proper interaction between thyroid hormone function and adrenal hormone function has physiologic consequences that affect our ability to forage (simulated by fastercise).[4] Proper thyroid system function and normal body temperature surely enhance the effects of fastercise (which releases epinephrine), and fastercise may enhance T4-to-T3 conversion. Also in chapter 7, I discussed the importance of the high concentration of mitochondria in brown adipose tissue (BAT), which promotes burning fat and generating heat. A 2018 study in the medical journal *Autophagy* showed that T3 activates BAT by increasing the destruction of worn-out mitochondria, increasing the production of new mitochondria, and increasing mitochondrial activity.[5] There's a parallel with high-intensity exercise like fastercise, because it also increases mitochondrial activity by increasing the size and number of mitochondria. The authors of the BAT study concluded that stimulation of BAT activity by thyroid hormone might represent a therapeutic strategy for obesity and metabolic diseases, and I agree with their conclusion.

A slow metabolism can surely contribute to easy weight gain, and correcting a slow metabolism can surely help alleviate it. T3 therapy is a treatment for normalizing low body temperature, or in other words, normalizing metabolic rate. I've seen patients lose over one hundred pounds while on T3 therapy when they were unable to lose weight while they were experiencing persistent low body temperature. One dramatic case was a woman who lost thirty-four pounds in a month, while changing nothing else in her diet or lifestyle. The only thing she reported doing differently was the T3 therapy. Weight-loss management is a very complicated issue, however, and normalizing body temperatures with T3 therapy has unpredictable results with regard to weight management. Sometimes, T3 therapy allows people to normalize their body temperature and relieve many symptoms, but they still have trouble losing weight. I believe fastercise holds promise in helping many of those people.

Just as T3 therapy involves cycling people on and off a program of thyroid hormones, fastercise generates a cycling of hormones as well. When we fastercise, our epinephrine, cortisol, glucagon, and growth hormone levels increase for a time and then gradually decrease. Since epinephrine promotes T4-to-T3 conversion, high-intensity exercise can be thought of as a small cycle of T3 as well. (Incidentally, eating a meal will also increase T3 production.) The effects of fastercise often seem to be cumulative. For example, notice how my body temperature rose higher with each successive bout of shivercise (see figure 10.3 on page 195). It may be that the more we send forage signals to the body by canceling hunger with fastercise, fastercising immediately before we eat and eating lightly, the clearer the message we send to the body and the faster our metabolism becomes. Average body temperature may be an indication of our current survival balance set point. (The lower the temperature, the more we're shifted toward storage mode.) In other words, Wilson's temperature syndrome may simply be the condition of being in storage mode. As mentioned above, WTS may be epidemic now partly because of the widespread convenience that has been afforded to us by modern technology. Shifting the survival balance out of storage mode and toward a normal temperature with fastercise and meal swinging may help people lose excess fat and, more important, keep it off. In my experience, people are much more likely to keep off body fat when they lose that excess fat while their body temperature is averaging in a normal range as compared to losing fat with a low temperature.

Summing Up

I hope the information in this book serves you well. In closing, I want to recap some of the most critical points about the power of fastercise. Fastercise is a very simple, time-efficient, natural, and sustainable form of exercise that also carries a low perception of difficulty. (It feels easy compared to other forms of exercise.) It can actually save you time and money by quickly reducing your hunger, thereby reducing the amount of time and money you need to spend on activities related to eating, such as grocery shopping, cooking, and doing dishes. Fastercise provides quick energy to our brains in the form of lactate and glucose. This quick energy can increase our mental focus, sharpen our thinking, and increase the productivity of our workdays. High-intensity exercise like fastercise stimulates nerve growth factor, which can aid in learning and reduce mental errors.

Rather than being unpleasant, fastercise actually feels good and relieves stress. Fastercise promotes the formation of endorphins (which reduce anxiety, tension, anger, and confusion) as well as endocannabinoids (which reduce pain and anxiety and generate a sense of well-being). Fastercise can stimulate the body's production of growth hormone and lactate, which can tighten the skin for a healthier, more youthful appearance and promote wound healing. Fastercise can greatly promote metabolic health and well-being by decreasing insulin resistance and increasing the size and number of our mitochondria, the cellular machinery that turns fat into energy. Improved energy processing may result in easier movement, better immune function, faster healing, more refreshing sleep, and fewer sick days. In addition, high-intensity exercise can decrease inflammation and autoimmune disease.

Fastercise can help provide remarkable improvement in athletic performance. Better metabolic performance also equates to better respiratory fitness. Fastercise may be significantly helpful for people with lung disease. Plus, almost anyone can easily fastercise, including people with physical limitations due to arthritis, old age, recent childbirth, blindness, or amputations. Thus, fastercise can help a wide range of people achieve better health by getting rid of excess fat, but it's not just about fat loss. Fastercise may help people recover from WTS, and it provides numerous benefits even for people who aren't obese. The power of fastercise is that it is a surprisingly pleasant, convenient, and effective way to get extremely fit in just a few minutes a day. Better fitness means being able to better enjoy time with our family and friends, make memories, and be more successful in almost every aspect of our lives. Happy fastercising!

ACKNOWLEDGMENTS

A special thank you to my colleague Michaël Friedman, whose friendship has been a tremendous blessing to me both personally and professionally. I greatly admire and appreciate his ability to turn a vision into reality step by step, and his proactive "Why not?" approach to living life. For nearly two decades he has been a staunch supporter of my work with Wilson's temperature syndrome and has helped bring about the certification training of hundreds of physicians. I'm grateful to him for connecting me with Chelsea Green Publishing (CGP) and for publisher Margo Baldwin's willingness to publish this book. I've greatly enjoyed working with my editor, Fern Marshall Bradley. I have been amazed at her ability to organize ideas and her obvious expertise in using words to brighten the lives of others. Her work has surely made this a much better book. I also appreciate the marvelous job the CGP production team has done in making this book look so great.

Sometimes, the best way to formalize an idea is by explaining it to someone else. Many thanks to my wife, Lisa, who has provided the invaluable contribution of listening to my thoughts in their most unrefined form and in every conceivable variation and iteration since then. Thanks so much for helping this book make sense.

Talk about an early adopter, my daughter Allison Roberts was the first person other than myself to fully implement the ideas presented in this book. Her discipline and diligence have enabled her to demonstrate the amazing results that are possible with this program. She has helped many people to implement this program in everyday life. Her passion for these principles helped to shape and bring forth *The Power of Fastercise* that you now hold in your hands. Thank you, Allison.

I feel we all owe a tremendous debt of gratitude to the computer science pioneers who laid the foundation for readily accessible information through the internet. This access to information is enabling us to better understand almost every aspect of our lives. I also gratefully acknowledge the pioneering work of medical researchers who provide pieces to the puzzle of life. Their broad shoulders provide a platform on which we stand to get a better view than we could ever get on our own. Specifically, I'd like to thank the researchers I have referenced in this book.

I express my gratitude to Tim Newcomb for the wonderful job he did illustrating this book and his amazing ability to meet deadlines. I'm very grateful also to those physicians and patients who listened to my explanations and implemented my ideas and provided much valuable confirmation and feedback. Finally, I'd like to thank my friends and colleagues at Restorative Formulations and the Association for the Advancement of Restorative Medicine for their many years of kindness and support that have made many aspects of my personal and professional life possible.

NOTES

Introduction

1. Stuart Flint et al., "Obesity Discrimination in the Recruitment Process: 'You're Not Hired!'" *Frontiers in Psychology* 7 (2016): 647, https://doi.org/10.3389/fpsyg.2016.00647.

Chapter 1: The New Science of Signaling Exercise

1. For example, Jameason Cameron et al., "Fasting for 24 Hours Heightens Reward from Food and Food-Related Cues," *PLoS ONE* 9, no. 1 (2014): e85970, https://doi.org/10.1371/journal.pone.0085970.

Chapter 2: It's All About Energy

1. George A. Brooks, "Cell-Cell and Intracellular Lactate Shuttles," *Journal of Physiology* 587, no. 23 (2009): 5591–600, https://doi.org/10.1113/jphysiol.2009.178350.
2. Jeremy M. Berg, John L. Tymoczko, and Lubert Stryer, *Biochemistry*, 5th ed. (New York: W. H. Freeman, 2002), Section 30.2, https://www.ncbi.nlm.nih.gov/books/NBK22436/.
3. Helene Nørrelund, "The Metabolic Role of Growth Hormone in Humans with Particular Reference to Fasting," *Growth Hormone and IGF Research* 15 (March 2005): 95–122, https://doi.org/10.1016/j.ghir.2005.02.005.
4. William McArdle, Frank Katch, and Victor Katch, *Essentials of Exercise Physiology*, 5th ed. (Philadelphia: Wolters Kluwer, 2015), 148.
5. Andrew Halestrap and Nigel Price, "The Proton-Linked Monocarboxylate Transporter (MCT) Family: Structure, Function and Regulation," *Biochemical Journal* 343 (1999): 296, https://doi.org/10.1042/bj3430281.
6. Richard Veech, "The Therapeutic Implications of Ketone Bodies: The Effects of Ketone Bodies in Pathological Conditions: Ketosis, Ketogenic Diet, Redox States, Insulin Resistance, and Mitochondrial Metabolism," *Prostaglandins, Leukotrienes and Essential Fatty Acids* 70, no. 3 (2004): 309–19, https://doi.org/10.1016/j.plefa.2003.09.007.
7. Melissa Ouellet et al., "Diffusion of Docosahexaenoic and Eicosapentaenoic Acids through the Blood-Brain Barrier: An *In Situ* Cerebral Perfusion Study," *Neurochemistry International* 55 (December 2009): 476–482, https://doi.org/10.1016/j.neuint.2009.04.018; Quentin R. Smith and Hiroshi Nagura, "Fatty Acid Uptake and Incorporation in Brain: Studies with the Perfusion Model," *Journal of Molecular Neuroscience* 16 (June 2001): 167–72, https://doi.org/10.1385/JMN:16:2-3:167; Reynold Spector, "Fatty Acid Transport through the Blood-Brain Barrier," *Journal of Neurochemistry* 50 (February

1988): 639–43, https://doi.org/10.1111/j.1471-4159.1988.tb02958.x; Wesley M. Williams et al., "*In Vivo* Incorporation from Plasma of Radiolabeled Palmitate and Arachidonate into Rat Brain Microvessels," *Microvascular Research* 53 (March 1997): 163–66, https://doi.org/10.1006/mvre.1996.1984.

8. Peter Schönfeld and Georg Reiser, "Why Does Brain Metabolism Not Favor Burning of Fatty Acids to Provide Energy?—Reflections on Disadvantages of the Use of Free Fatty Acids as Fuel for Brain," *Journal of Cerebral Blood Flow and Metabolism* 33, no. 10 (October 2013): 1493, https://doi.org/10.1038/jcbfm.2013.128.

9. George Cahill, "Fuel Metabolism in Starvation," *Annual Review of Nutrition* 26, no. 22 (2006): 10, https://doi.org/10.1146/annurev.nutr.26.061505.111258; Michael Stumvoll et al., "Role of Glutamine in Human Carbohydrate Metabolism in Kidney and Other Tissues," *Kidney International* 55, no. 3 (March 1999): 786, https://doi.org/10.1046/j.1523-1755.1999.055003778.x.

10. Gerald Dienel, "Brain Lactate Metabolism: The Discoveries and the Controversies," *Journal of Cerebral Blood Flow and Metabolism* 32, no. 7 (July 2012): 1107, https://doi.org/10.1038/jcbfm.2011.175.

11. Curtis Triplitt, "Understanding the Kidney's Role in Blood Glucose Regulation," *American Journal of Managed Care* 18 (April 2012): S11.

12. Joseph Katz and John A. Tayek, "Gluconeogenesis and the Cori Cycle in 12-, 20-, and 40-h-Fasted Humans," *American Journal of Physiology Endocrinology and Metabolism* 275, no. 3 (September 1998): E537–42. https://doi.org/10.1152/ajpendo.1998.275.3.E373.

13. Cahill, "Fuel Metabolism," 6.

14. Jeff Volek et al., "Metabolic Characteristics of Keto-Adapted Ultra-Endurance Runners," *Metabolism* 65, no. 3 (March 2016): 100–110, https://doi.org/10.1016/j.metabol.2015.10.028.

Chapter 3: Hormones Send Key Signals

1. Trevor Lamb and Edward Pugh, Jr., "Phototransduction, Dark Adaptation, and Rhodopsin Regeneration," *Investigative Ophthalmology and Visual Science* 47, no. 12 (2006): 5138–52, https://doi.org/10.1167/iovs.06-0849.

2. Jason Fung, *The Obesity Code: Unlocking the Secrets of Weight Loss* (Vancouver: Greystone Books, 2016).

3. Robert Henry et al., "Intensive Conventional Insulin Therapy for Type II Diabetes," *Diabetes Care* 16, no. 1 (January 1993): 23–31, https://doi.org/10.2337/diacare.16.1.21.

4. Ling Chun Kong et al., "Insulin Resistance and Inflammation Predict Kinetic Body Weight Changes in Response to Dietary Weight Loss and Maintenance in Overweight and Obese Subjects by Using a Bayesian Network Approach," *American Journal of Clinical Nutrition* 98, no. 6 (December 2013): 1385–94, https://doi.org/10.3945/ajcn.113.058099.

5. Gianni Biolo et al., "Insulin Action on Muscle Protein Kinetics and Amino Acid Transport during Recovery after Resistance Exercise," *Diabetes* 48, no. 5 (1999):

949–57, https://doi.org/10.2337/diabetes.48.5.949; Satoshi Fujita et al., "Effect of Insulin on Human Skeletal Muscle Protein Synthesis Is Modulated by Insulin-Induced Changes in Muscle Blood Flow and Amino Acid Availability," *American Journal of Physiology-Endocrinology and Metabolism* 291, no. 4 (October 2006): E745–54, https://doi.org/10.1152/ajpendo.00271.2005; Haitham Abdulla et al., "Role of Insulin in the Regulation of Human Skeletal Muscle Protein Synthesis and Breakdown: A Systematic Review and Meta-Analysis," *Diabetologia* 59, no. 1 (January 2016): 44–55, https://doi.org/10.1007/s00125-015-3751-0; Giuseppe Caso and Margaret McNurlan, "Effect of Insulin on Whole Body Protein Metabolism in Children with Type 1 Diabetes," *Current Opinion in Clinical Nutrition and Metabolic Care* 13, no. 1 (2010): 93–96, https://doi.org/10.1097/mco.0b013e328333294d.

6. Xiaonan Wang et al., "Insulin Resistance Accelerates Muscle Protein Degradation: Activation of the Ubiquitin-Proteasome Pathway by Defects in Muscle Cell Signaling," *Endocrinology* 147, no. 9 (September 2006): 4160–68, https://doi.org/10.1210/en.2006-0251; Christelle Guillet and Yves Boirie, "Insulin Resistance: A Contributing Factor to Age-Related Muscle Mass Loss?" *Diabetes and Metabolism* 31, spec no. 2 (December 2005): 5S20–26, https://doi.org/10.1016/s1262-3636(05)73648-x.

7. Alfredo Quiñones-Galvan and Ele Ferrannini, "Renal Effects of Insulin in Man," *Journal of Nephrology* 10, no. 4 (July–August 1997): 188–91.

8. James Conner et al., "NGF Is Essential for Hippocampal Plasticity and Learning," *Journal of Neuroscience* 29, no. 35 (September 2009): 10883–89, https://doi.org/10.1523/jneurosci.2594-09.2009.

9. Zhengtang Qi and Shuzhe Ding, "Obesity-Associated Sympathetic Overactivity in Children and Adolescents: The Role of Catecholamine Resistance in Lipid Metabolism," *Journal of Pediatric Endocrinology and Metabolism* 29, no. 2 (2016): 113–25, https://doi.org/10.1515/jpem-2015-0182.

10. Chunli Yu et al., "Mechanism by Which Fatty Acids Inhibit Insulin Activation of Insulin Receptor Substrate-1 (IRS-1)-Associated Phosphatidylinositol 3-Kinase Activity in Muscle," *Journal of Biological Chemistry* 277, no. 52 (December 2002): 50230–36, https://doi.org/10.1074/jbc.M200958200; Helene Nørrelund et al., "The Decisive Role of Free Fatty Acids for Protein Conservation during Fasting in Humans with and without Growth Hormone," *Journal of Clinical Endocrinology and Metabolism* 88, no. 9 (2003): 4371–78, https://doi.org/10.1210/jc.2003-030267.

11. Helene Nørrelund, Anne Riis, and Niels Møller, "Effects of GH on Protein Metabolism during Dietary Restriction in Man," *Growth Hormone and IGF Research* 12, no. 4 (2002): 198–207, https://doi.org/10.1016/s1096-6374(02)00043-6.

12. Nørrelund, "The Metabolic Role."

13. Nørrelund, "The Metabolic Role."

14. Åsa Tivesten et al., "Growth Hormone-Induced Blood Pressure Decrease Is Associated with Increased mRNA Levels of the Vascular Smooth Muscle KATP Channel," *Journal of Endocrinology* 183, no. 1 (October 2004): 195–202.

15. Guilherme Póvoa and Lucia Diniz, "Growth Hormone System: Skin Interactions," *Anais Brasileiros de Dermatologia* 86, no. 6 (2011): 1159–65, https://doi.org/10.1590/s0365-05962011000600015.

16. Póvoa and Diniz, "Growth Hormone."

17. Nørrelund, "The Metabolic Role."

18. Jen-Chieh Chuang et al., "Ghrelin Directly Stimulates Glucagon Secretion from Pancreatic α-Cells," *Molecular Endocrinology* 25, no. 9 (September 2011): 1600–1611, https://doi.org/10.1210/me.2011-1001.

19. James King et al., "Exercise and Ghrelin. A Narrative Overview of Research," *Appetite* 68 (2013): 83–91, https://doi.org/10.1016/j.appet.2013.04.018.

20. Joëlle Dupont et al., "Ghrelin in Female and Male Reproduction," *International Journal of Peptides* 2010 (2010): 158102, https://doi.org/10.1155/2010/158102.

21. C. B. Yang and C. C. Chuang, "Effects of an Acute Bout of Exercise on Serum Soluble Leptin Receptor (sOB-R) Levels," *Journal of Sports Sciences* 32, no. 5 (2014): 446–51, http://doi.org/ 10.1080/02640414.2013.828848.

22. McArdle, Katch, and Katch, *Essentials*, 54.

23. Josefine Lindroos et al., "Human but Not Mouse Adipogenesis Is Critically Dependent on LMO3," *Cell Metabolism* 18, no. 1 (2013): 62–74, https://doi.org/10.1016/j.cmet.2013.05.020.

24. Mi-Jeong Lee et al., "Deconstructing the Roles of Glucocorticoids in Adipose Tissue Biology and the Development of Central Obesity," *Biochimica et Biophysica Acta* 1842, no. 3 (March 2014): 473–81, https://doi.org/10.1016/j.bbadis.2013.05.029.

25. H. Galbo, J. Holst, and N. Christensen, "Glucagon and Plasma Catecholamine Responses to Graded and Prolonged Exercise in Man," *Journal of Applied Physiology* 38, no. 1 (January 1975): 70–76, https://doi.org/10.1152/jappl.1975.38.1.70; C. Lavoie et al., "Glucose Metabolism during Exercise in Man: The Role of Insulin and Glucagon in the Regulation of Hepatic Glucose Production and Gluconeogenesis," *Canadian Journal of Physiology and Pharmacology* 75, no. 1 (January 1997): 26–35, https://doi.org/10.1139/y96-161.

Chapter 4: The Non-Fed State

1. Andreas Michalsen and Chenying Li, "Fasting Therapy for Treating and Preventing Disease—Current State of Evidence," *Forsch Komplementmed* 20, no. 6 (2013): 444–53, https://doi.org/10.1159/000357765.

2. Shigetada Furukawa et al., "Increased Oxidative Stress in Obesity and Its Impact on Metabolic Syndrome," *Journal of Clinical Investigation* 114, no. 12 (December 2004): 1752–61, https://doi.org/10.1172/jci21625.

3. Kim Stote et al., "A Controlled Trial of Reduced Meal Frequency without Caloric Restriction in Healthy, Normal-Weight, Middle-Aged Adults," *American Journal of Clinical Nutrition* 85, no. 4 (April 2007): 981–88, https://doi.org/10.1093/ajcn/85.4.981.

4. Robert Henry et al., "Intensive Conventional Insulin Therapy for Type II Diabetes: Metabolic Effects during a 6-Mo Outpatient Trial," *Diabetes Care* 16, no. 1 (January 1993): 21–31, https://doi.org/10.2337/diacare.16.1.21.

5. Eric Jéquier, "Energy Expenditure in Obesity," *Clinics in Endocrinology and Metabolism* 13, no. 3 (November 1984): 563–80, https://doi.org/10.1016/S0300-595X(84)80038-9.

6. Jameason Cameron et al., "Fasting for 24 Hours Heightens Reward from Food and Food-Related Cues," *PLoS ONE* 9, no. 1 (2014): e85970, https://doi.org/10.1371/jounal.pone.0085970.

7. Scott Powers and Edward Howley, *Exercise Physiology* (New York: McGraw-Hill, 2011).

8. Rudolph Leibel, "Molecular Physiology of Weight Regulation in Mice and Humans," *International Journal of Obesity 32* (2008): S98–S99, https://doi.org/10.1038/ijo2008245.

9. Fung, *Obesity Code*.

10. W. Larry Kenney, *Physiology of Sport and Exercise*, 5th ed. (Champaign: Human Kinetics, 2018).

11. Ancel Keys et al., *The Biology of Human Starvation* (Minneapolis: University of Minnesota Press, 1950).

12. Michael Rosenbaum et al., "Long-Term Persistence of Adaptive Thermogenesis in Subjects Who Have Maintained a Reduced Body Weight," *American Journal of Clinical Nutrition* 88, no. 4 (October 2008): 906–12, https://doi.org/10.1093/ajcn/88.4.906.

13. Katie Mason et al., *Overview of Gut Immunology* (New York: Springer-Verlag, 2008).

14. K. A. Varady, "Intermittent versus Daily Calorie Restriction: Which Diet Regimen Is More Effective for Weight Loss?," *Obesity Reviews* 12, no. 7 (July 2011): e593–601, https://doi.org/10.1111/j.1467-789x.2011.00873.x.

15. Monica Klempel et al., "Dietary and Physical Activity Adaptations to Alternate Day Modified Fasting: Implications for Optimal Weight Loss," *Nutrition Journal* 9, no. 1 (2010): 35, https://doi.org/10.1186/1475-2891-9-35.

16. P. Mansell, I. Fellows, and I. Macdonald, "Enhanced Thermogenic Response to Epinephrine after 48-h Starvation in Humans," *American Journal of Physiology-Regulatory, Integrative and Comparative Physiology* 258, no. 1 (1990): R87–93, https://doi.org/10.1152/ajpregu.1990.258.1.r87.

17. Mehrdad Alirezaei, "Short-Term Fasting Induces Profound Neuronal Autophagy," *Autophagy* 6, no. 6 (August 2010): 702–10, https://doi.org/10.4161/auto.6.6.12376.

18. Bronwen Martin, Mark Mattson, and Stuart Maudsley, "Caloric Restriction and Intermittent Fasting: Two Potential Diets for Successful Brain Aging," *Ageing Research Reviews* 5, no. 3 (August 2006): 332–53, https://doi.org/10.1016/j.arr.2006.04.002; Mark Mattson, "Energy Intake, Meal Frequency, and Health: A Neurobiological Perspective," *Annual Review of Nutrition* 25, no. 1 (2005): 237–60, https://doi.org/10.1146/annurev.nutr.25.050304.092526; M. Tajes et al., "Neuroprotective Role of Intermittent Fasting in Senescence-Accelerated Mice P8 (SAMP8)," *Experimental Gerontology* 45, no. 9 (September 2010): 702–10, https://doi.org/10.1016/jexger.2010.04.010.

19. Mattson, "Energy Intake."

20. E. Wright, J. Scism-Bacon, and L. Glass, "Oxidative Stress in Type 2 Diabetes: The Role of Fasting and Postprandial Glycaemia," *International Journal of Clinical Practice* 60, no. 3 (March 2006): 308–14, https://doi.org/10.1111/j.1368-5031.2006.00825.x.

21. Mo'ez Al-Islam Ezzat Faris et al., "Intermittent Fasting during Ramadan Attenuates Proinflammatory Cytokines and Immune Cells in Healthy Subjects," *Nutrition Research* 32, no. 12 (December 2012): 947–55, https://doi.org/10.1016/j.nutres.2012.06.021.

22. Chia-Wei Cheng et al., "Prolonged Fasting Reduces IGF-1/PKA to Promote Hematopoietic-Stem-Cell-Based Regeneration and Reverse Immunosuppression," *Cell Stem Cell* 14, no. 6 (June 2014): 810–23, https://doi.org/10.1016/j.stem.2014.04.014.

23. James Brown, Michael Mosley, and Sarah Aldred, "Intermittent Fasting: A Dietary Intervention for Prevention of Diabetes and Cardiovascular Disease?" *The British Journal of Diabetes and Vascular Disease* 13, no. 2 (2013): 68–72, https://doi.org/10.1177/1474651413486496.

24. Brown, Mosley, and Aldred, "Intermittent Fasting"; L. Dewanti et al., "Unexpected Changes in Blood Pressure and Hematological Parameters among Fasting and Nonfasting Workers during Ramadan in Indonesia," *European Journal of Clinical Nutrition* 60, no. 7 (July 2006): 877–81, https://doi.org/10.1038/sj.ejcn.1602393; Luigi Fontana et al., "Calorie Restriction or Exercise: Effects on Coronary Heart Disease Risk Factors. A Randomized, Controlled Trial," *American Journal of Physiology–Endocrinology and Metabolism* 293, no. 1 (July 2007): E197–202, https://doi.org/10.1152/ajpendo.00102.2007.

25. Alan Goldhamer et al., "Medically Supervised Water-Only Fasting in the Treatment of Hypertension," *Journal of Manipulative and Physiological Therapeutics* 24, no. 5 (June 2001) 335–339, https://doi.org/10.1067/mmt.2001.115263.

26. Jajesh Katare, "Chronic Intermittent Fasting Improves the Survival Following Large Myocardial Ischemia by Activation of BDNF/VEGF/PI3K Signaling Pathway," *Journal of Molecular and Cellular Cardiology* 46, no. 3 (2009): 405–12, https://doi.org/10.1016/j.yjmcc.2008.10.027.

27. Anton Carlson and Frederick Hoelzel, "Apparent Prolongation of the Life Span of Rats by Intermittent Fasting," *Journal of Nutrition* 32, no. 3 (March 1946): 363–75, https://doi.org/10.1093/jn/31.3.363.

28. Carlson and Hoelzel, "Apparent Prolongation."

29. Heather Weir et al., "Dietary Restriction and AMPK Increase Lifespan via Mitochondrial Network and Peroxisome Remodeling," *Cell Metabolism* 26, no. 6 (December 2017): 884–96.e5, https://doi.org/10.1016/j.cmet.2017.09.024.

30. S. Herzig and R. J. Shaw, "AMPK: Guardian of Metabolism and Mitochondrial Homeostasis," *Nature Reviews Molecular Cell Biology* 19 (2018): 121–35, https://doi.org/10.1038/nrm.2017.95.

31. Antero Salminen and Kai Kaarniranta, "AMP-Activated Protein Kinase (AMPK) Controls the Aging Process via an Integrated Signaling Network," *Ageing Research Reviews* 11, no. 2 (2012): 230–34, https://doi.org/10.1016/j.arr.2011.12.005.

32. Nuria Martinez-Lopez, Diana Athonvarangkul, and Rajat Singh, "Autophagy and Aging," *Advances in Experimental Medicine and Biology* 847 (2015): 73–87, https://doi.org/10.1007/978-1-4939-2404-2_3.

33. Vladimir Anisimov, "Insulin/IGF-1 Signaling Pathway Driving Aging and Cancer as a Target for Pharmacological Intervention," *Experimental Gerontology* 38, no. 10 (October 2003): 1041–49, https://doi.org/10.1016/s0531-5565(03)00169-4.

Chapter 5: Finding the Right Balance

1. Rudolph Leibel, Michael Rosenbaum, and Jules Hirsch, "Changes in Energy Expenditure Resulting from Altered Body Weight," *New England Journal of Medicine* 332, no. 10 (March 1995): 621–28, https://doi.org/10.1056/nejm 199503093321001.
2. Erik Diaz et al., "Metabolic Response to Experimental Overfeeding in Lean and Overweight Healthy Volunteers," *American Journal of Clinical Nutrition* 56, no. 4 (October 1992): 541–55, https://doi.org/10.1093/ajcn/56.4.641.
3. J. Webber and I. Macdonald, "The Cardiovascular, Metabolic and Hormonal Changes Accompanying Acute Starvation in Men and Women," *British Journal of Nutrition* 71, no. 03 (March 1994): 437–47, https://doi.org/10.1079/bjn19940150.
4. Christian Zauner et al., "Resting Energy Expenditure in Short-Term Starvation Is Increased as a Result of an Increase in Serum Norepinephrine," *American Journal of Clinical Nutrition* 71, no. 6 (June 2000): 1511–15, https://doi.org/10.1093/ajcn/71.6.1511.
5. Fung, *Obesity Code*.
6. Fung, *Obesity Code*.
7. Priya Sumithran et al., "Long-Term Persistence of Hormonal Adaptations to Weight Loss," *New England Journal of Medicine* 365, no. 17 (October 2011): 1597–604, https://doi.org/10.1097/ogx.0b013e318247c6f7.
8. Fung, *Obesity Code*.

Chapter 6: Getting the Most from Fastercise

1. Rena Wing, "Physical Activity in the Treatment of the Adulthood Overweight and Obesity: Current Evidence and Research Issues," *Medicine and Science in Sports and Exercise* 31, no. 11 supplement (November 1999): S547–52, https://doi.org/10.1097 /00005768-199911001-00010.
2. McArdle, Katch, and Katch, *Essentials*, 148.
3. Jergen Jensen et al., "The Role of Skeletal Muscle Glycogen Breakdown for Regulation of Insulin Sensitivity by Exercise," *Frontiers in Physiology* 2 (2011): 112, https://doi .org/10.3389/fphys.2011.00112.
4. I. De Glisezinski et al., "Lack of α2-Adrenergic Antilipolytic Effect during Exercise in Subcutaneous Adipose Tissue of Trained Men," *Journal of Applied Physiology* 91, no. 4 (2001): 1760–65, https://doi.org/10.1152/jappl.2001.91.4.1760.
5. Chris McGlory and Stuart Phillips, "Exercise and the Regulation of Skeletal Muscle Hypertrophy," *Molecular and Cellular Regulation of Adaptation to Exercise* (2015): 153–73, https://doi.org/10.1016/bs.pmbts.2015.06.018.
6. McGlory and Phillips, "Exercise."
7. McArdle, Katch, and Katch, *Essentials*, 115–17.

8. E. Trapp et al., "The Effects of High-Intensity Intermittent Exercise Training on Fat Loss and Fasting Insulin Levels of Young Women," *International Journal of Obesity* 32, no. 4 (April 2008): 684–91, https://doi.org/10.1038/sj.ijo.0803781.

9. Hiroto Sasaki et al., "4 Weeks of High-Intensity Interval Training Does Not Alter the Exercise-Induced Growth Hormone Response in Sedentary Men," *Springerplus* 3, no. 1 (2014): 336, https://doi.org/10.1186/2193-1801-3-336.

10. Póvoa and Diniz, "Growth Hormone."

11. Póvoa and Diniz, "Growth Hormone."

12. Trapp et al., "The Effects of High-Intensity."

13. G. Jones, "Caffeine and Other Sympathomimetic Stimulants: Modes of Action and Effects on Sports Performance," *Essays Biochemistry* 44 (2008): 109–23, https://doi.org/10.1042/BSE0440109..

14. Daniel Pink, *When: The Scientific Secrets of Perfect Timing* (New York: Penguin Random House, 2018).

15. Alex Wadley et al., "Low Volume-High Intensity Interval Exercise Elicits Antioxidant and Anti-Inflammatory Effects in Humans," *Journal of Sports Sciences* 34, no. 1 (2016): 1–9, https://doi.org/10.1080/02640414.2015.1035666; Anne Petersen and Bente Pedersen, "The Anti-Inflammatory Effect of Exercise," *Journal of Applied Physiology* 98, no. 4 (April 2005): 1154–62, https://doi.org/10.1152/japplphysiol.00164.2004.

16. McArdle, Katch, and Katch, *Essentials*, 392.

17. McArdle, Katch, and Katch, 594.

18. Jeffrey Metter et al., "Skeletal Muscle Strength as a Predictor of All-Cause Mortality in Healthy Men," *Journals of Gerontology Series A: Biological Sciences and Medical Sciences* 57, no. 10 (October 2002): B359–65, https://doi.org/10.1093/gerona/57.10.b359.

Chapter 7: The Physiology of Fastercise

1. McArdle, Katch, and Katch, *Essentials*, 392.

2. McArdle, Katch, and Katch, 166.

3. Douglas Ballor et al., "Resistance Weight Training during Caloric Restriction Enhances Lean Body Weight Maintenance," *American Journal of Clinical Nutrition* 47, no. 1 (1988): 19–25, https://doi.org/10.1093/ajcn/47.1.19.

4. Edward Weiss et al., "Lower Extremity Muscle Size and Strength and Aerobic Capacity Decrease with Caloric Restriction but Not with Exercise-Induced Weight Loss," *Journal of Applied Physiology* 102, no. 2 (2007): 634–40, https://doi.org/10.1152/japplphysiol.00853.2006.

5. Karen Van Proeyen et al., "Beneficial Metabolic Adaptations due to Endurance Exercise Training in the Fasted State," *Journal of Applied Physiology* 110, no. 1 (January 2011): 236–45, https://doi.org/10.1152/japplphysiol.00907.2010.

6. McArdle, Katch, and Katch, *Essentials*, 393.

7. Herman Pontzer et al., "Constrained Total Energy Expenditure and Metabolic Adaptation to Physical Activity in Adult Humans," *Current Biology* 26, no. 3 (2016):

410–17, https://doi.org/10.1016/j.cub.2015.12.046; Herman Pontzer, "The Exercise Paradox," *Scientific American*, February 2017.

8. McArdle, Katch, and Katch, *Essentials*, 414.

9. McArdle, Katch, and Katch, 423.

10. Robert Jacobs et al., "Improvements in Exercise Performance with High-Intensity Interval Training Coincide with an Increase in Skeletal Muscle Mitochondrial Content and Function," *Journal of Applied Physiology* 115, no. 6 (September 2013): 785–93, https://doi.org/10.1152/japplphysiol.00445.2013.

11. McArdle, Katch, and Katch, *Essentials*, 425; Martin Gibala et al., "Physiological Adaptations to Low-Volume, High-Intensity Interval Training in Health and Disease," *Journal of Physiology* 590, no. 5 (2012): 1077–84, https://doi.org/10.1113/jphysiol.2011.224725.

12. Wing, "Physical Activity."

13. Gerrit van Hall, "The Physiological Regulation of Skeletal Muscle Fatty Acid Supply and Oxidation During Moderate-Intensity Exercise," *Sports Medicine* 45, no. S1 (2015): 23–32, https://doi.org/10.1007/s40279-015-0394-8.

14. Aaron Sim et al., "High-Intensity Intermittent Exercise Attenuates ad-Libitum Energy Intake," *International Journal of Obesity* 38, no. 3 (March 2014): 417–22, https://doi.org/10.1038/ijo.2013.102; Jessica Douglas et al., "Acute Exercise and Appetite-Regulating Hormones in Overweight and Obese Individuals: A Meta-Analysis," *Journal of Obesity* (2016): 1–8, https://doi.org/10.1155/2016/26436251; Matthew Schubert et al., "Acute Exercise and Hormones Related to Appetite Regulation: A Meta-Analysis," *Sports Medicine* 44, no. 3 (March 2014): 387–403, https://doi.org/10.1007/s40279-013-0120-3.

15. Aaron Sim et al., "Effects of High-Intensity Intermittent Exercise Training on Appetite Regulation," *Medicine and Science in Sports and Exercise* 47, no. 11 (November 2015): 2441–49, https://doi.org/10.1249/mss.0000000000000687.

16. McArdle, Katch, and Katch, *Essentials*, 398.

17. P. Sparling et al., "Exercise Activates the Endocannabinoid System," *NeuroReport* 14, no. 7 (December 2003): 2209–11, https://doi.org/10.1097/00001756-200312020-00015.

18. Sarah Dubreucq et al., "CB1 Receptor Deficiency Decreases Wheel-Running Activity: Consequences on Emotional Behaviors and Hippocampal Neurogenesis," *Experimental Neurology* 224, no. 1 (July 2010): 106–13, https://doi.org/10.1016/j.expneurol.2010.01.017.

19. Brunella Capaldo et al., "Epinephrine Directly Antagonizes Insulin-Mediated Activation of Glucose Uptake and Inhibition of Free Fatty Acid Release in Forearm Tissues," *Metabolism* 41, no. 10 (October 1992): 1146–49, https://doi.org/10.1016/0026-0495(92)90301-p; Bernard Leboeuf, Robert Flinn, and George Cahill, "Effect of Epinephrine on Glucose Uptake and Glycerol Release by Adipose Tissue in vitro," *Experimental Biology and Medicine* 102, no. 3 (1959): 527–29, https://doi.org/10.3181/00379727-102-25306; Bela Issekutz, "Role of Beta-Adrenergic Receptors in Mobilization of Energy Sources in Exercising Dogs," *Journal of Applied*

Physiology: Respiratory, Environmental and Exercise Physiology 44, no. 6 (June 1978): 869–76, https://doi.org/10.1152/jappl.1978.44.6.869.

20. Berg, Tymoczko, and Stryer, *Biochemistry*.

21. Hassan Zouhal et al., "Catecholamines and the Effects of Exercise, Training and Gender," *Sports Medicine* 38, no. 5 (2008): 401–23, https://doi.org/10.2165/00007256 -200838050-00004.

22. Zouhal et al., "Catecholamines."

23. Zouhal et al., "Catecholamines."

24. Zouhal et al., "Catecholamines."

25. Berg, Tymoczko, and Stryer, *Biochemistry*, section 21.3.1.

26. McArdle, Katch, and Katch, *Essentials*, 392.

27. Jeffrey Greiwe et al., "Norepinephrine Response to Exercise at the Same Relative Intensity before and after Endurance Exercise Training," *Journal of Applied Physiology* 86, no. 2 (1999): 531–35, https://doi.org/10.1152/jappl.1999.86.2.531.

28. Enrique Silva and Suzy Bianco, "Thyroid–Adrenergic Interactions: Physiological and Clinical Implications," *Thyroid* 18, no. 2 (2008): 157–65, https://doi.org/10.1089 /thy.2007.0252.

29. Silva and Bianco, "Thyroid–Adrenergic."

30. Federica Cioffi et al., "Thyroid Hormones and Mitochondria: With a Brief Look at Derivatives and Analogues," *Molecular and Cellular Endocrinology* 379, no. 1–2 (October 2013): 51–61, https://doi.org/10.1016/j.mce.2013.06.006; Sutapa Mukherjee et al., "Supplementation of T3 Recovers Hypothyroid Rat Liver Cells from Oxidatively Damaged Inner Mitochondrial Membrane Leading to Apoptosis," *BioMed Research International* 1–12 (2014): 590897, https://doi.org/10.1155/2014/590897.

31. Silva and Bianco, "Thyroid–Adrenergic"; Yasuhiro Hosoi et al., "Expression and Regulation of Type II Iodothyronine Deiodinase in Cultured Human Skeletal Muscle Cells," *Journal of Clinical Endocrinology and Metabolism* 84, no. 9 (1999): 3293–300, https://doi.org/10.1210/jcem.84.9.5969.

32. Silva and Bianco, "Thyroid–Adrenergic."

33. Yvonne Kilian et al., "Markers of Biological Stress in Response to a Single Session of High-Intensity Interval Training and High-Volume Training in Young Athletes," *European Journal of Applied Physiology* 116, no. 11–12 (December 2016): 2177–86, https://doi.org/10.1007/s00421-016-3467-y.

34. Alex Vermeulen, S. Goemaere, and Jean Kaufman, "Testosterone, Body Composition and Aging," *Journal of Endocrinological Investigation* 22, no. 5 supplement (1999): 110–16.

35. McArdle, Katch, and Katch, *Essentials*, 393.

36. Vermeulen, Goemaere, and Kaufman, "Testosterone."

37. Hans Kley et al., "Enhanced Conversion of Androstenedione to Estrogens in Obese Males," *Journal of Clinical Endocrinology and Metabolism* 51, no. 5 (November 1980): 1128–32, https://doi.org/10.1210/jcem-51-5-1128; Gladys Strain et al., "Effect of Massive Weight Loss on Hypothalamic-Pituitary-Gonadal Function in Obese Men,"

Journal of Clinical Endocrinology and Metabolism 66, no. 5 (May 1988): 1019–23, https://doi.org/10.1210/jcem-66-5-1019.

38. Alex Vermeulen, Martin Kaufman, and Vito Giagulli, "Influence of Some Biological Indexes on Sex Hormone-Binding Globulin and Androgen Levels in Aging or Obese Males," *Journal of Clinical Endocrinology and Metabolism* 81, no. 5 (May 1996): 1821–26, https://doi.org/10.1210/jcem.81.5.8626841.

39. Patrick Wahl et al., "Effect of High- and Low-Intensity Exercise and Metabolic Acidosis on Levels of GH, IGF-1, IGFBP-3 and Cortisol," *Growth Hormone and IGF Research* 20, no. 5 (2010): 380–85, https://doi.org/10.1016/j.ghir.2010.08.001.

40. Jonathan Peake et al., "Metabolic and Hormonal Responses to Isoenergetic High-Intensity Interval Exercise and Continuous Moderate-Intensity Exercise," *American Journal of Physiology-Endocrinology and Metabolism* 307, no. 7 (October 2014): E539–52, https://doi.org/10.1152/ajpendo.00276.2014.

41. Perveen Ghani, Silvia Wagner, and Zamirul Hussain, "Role of ADP-Ribosylation in Wound Repair. The Contributions of Thomas K. Hunt, MD," *Wound Repair and Regeneration* 11, no. 6 (2003): 439–44, https://doi.org/10.1046/j.1524-475x.2003.11608.x; Odilo Trabold et al., "Lactate and Oxygen Constitute a Fundamental Regulatory Mechanism in Wound Healing," *Wound Repair and Regeneration* 11, no. 6 (2003): 504–9, https://doi.org/10.1046/j.1524-475x.2003.11621.x; Patrick Wahl et al., "Effects of Acid-Base Balance and High or Low Intensity Exercise on VEGF and bFGF," *European Journal of Applied Physiology* 111, no. 7 (2010): 1405–13, https://doi.org/10.1007/s00421-010-1767-1.

42. James King et al., "Exercise and Ghrelin. A Narrative Overview of Research," *Appetite* 68 (2013): 83–91, https://doi.org/10.1016/j.appet.2013.04.018.

43. King et al., "Exercise and Ghrelin."

44. Johannes Erdmann et al., "Plasma Ghrelin Levels during Exercise—Effects of Intensity and Duration," *Regulatory Peptides* 143, no. 1–3 (October 2007): 127–35, https://doi.org/10.1016/j.regpep.2007.05.002; King et al., "Exercise and Ghrelin."

45. King et al., "Exercise and Ghrelin."

46. J. L. Goldstein et al., "Surviving Starvation: Essential Role of the Ghrelin-Growth Hormone Axis," *Cold Spring Harbor Symposia on Quantitative Biology* 76 (2011): 121–27, https://doi.org/10.1101/sqb.2011.76.010447.

47. King et al., "Exercise and Ghrelin."

48. Nørrelund, "The Metabolic Role."

49. King et al., "Exercise and Ghrelin."

50. King et al., "Exercise and Ghrelin."

51. King et al., "Exercise and Ghrelin."

52. Erdmann et al., "Plasma Ghrelin Levels."

53. Wahl, "Effect of High- and Low-Intensity Exercise."

54. McArdle, Katch, and Katch, *Essentials*, 378.

55. McArdle, Katch, and Katch, 378.

56. Yvonne Kilian et al., "Markers."

57. Robert Sherwin and Luigi Saccà, "Effect of Epinephrine on Glucose Metabolism in Humans: Contribution of the Liver," *American Journal of Physiology-Endocrinology and Metabolism* 247, no. 2 (August 1984): E157–65, https://doi.org/10.1152/ajpendo .1984.247.2.e157.

58. Wahl, "Effect of High- and Low-Intensity Exercise"; Hideo Takahashi et al., "Mechanism of Impaired Growth Hormone Secretion in Patients with Cushing's Disease," *European Journal of Endocrinology* 127, no. 1 (July 1992): 13–17, https://doi .org/10.1530/acta.0.1270013.

59. Dennis-Peter Born, Christoph Zinner, and Billy Sperlich, "The Mucosal Immune Function Is Not Compromised during a Period of High-Intensity Interval Training. Is It Time to Reconsider an Old Assumption?" *Frontiers in Physiology* 8 (2017): 485, https://doi.org/10.3389/fphys.2017.00485; David Nieman, "Risk of Upper Respiratory Tract Infection in Athletes: An Epidemiological and Immunologic Perspective," *Journal of Athletic Training* 32, no. 4 (October 1997): 344–49, https://doi .org/10.0000/www.ncbi.nlm.nih.gov/PMC1320353.

60. Stephen Boutcher, "High-Intensity Intermittent Exercise and Fat Loss," *Journal of Obesity* (2011): 868305, https://doi.org/10.1155/2011/868305.

61. Trapp et al., "The Effects of High-Intensity."

62. Stephan Nieuwoudt et al., "Functional High-Intensity Training Improves Pancreatic β-Cell Function in Adults with Type 2 Diabetes," *American Journal of Physiology-Endocrinology and Metabolism* 313, no. 3 (September 2017): E314–20, https://doi .org/10.1152/ajpendo.00407.2016; Tanja Sjöros et al., "Increased Insulin-Stimulated Glucose Uptake in Both Leg and Arm Muscles after Sprint Interval and Moderate Intensity Training in Subjects with Type 2 Diabetes or Prediabetes," *Scandinavian Journal of Medicine and Science in Sports* 28, no. 1 (January 2018): 77–87, https://doi .org/10.1111/sms.12875.

63. Erik Richter et al., "Effect of Exercise on Insulin Action in Human Skeletal Muscle," *Journal of Applied Physiology* 66, no. 2 (February 1989): 876–85, https://doi.org /10.1152/jappl.1989.66.2.876; John Devlin and Edward Horton, "Effects of Prior High-Intensity Exercise on Glucose Metabolism in Normal and Insulin-Resistant Men," *Diabetes* 34, no. 10 (October 1985): 973–79, https://doi.org/10.2337/diab .34.10.973; Ruth Burstein et al., "Effect of an Acute Bout of Exercise on Glucose Disposal in Human Obesity," *Journal of Applied Physiology* 69, no. 1 (July 1990): 299–304, https://doi.org/10.1152/jappl.1990.69.1.299; S. Bordenave et al., "Effects of Acute Exercise on Insulin Sensitivity, Glucose Effectiveness and Disposition Index in Type 2 Diabetic Patients," *Diabetes and Metabolism* 34, no. 3 (June 2008): 250–57, https://doi.org/10.1016/j.diabet.2007.12.008.

64. K. Sahlin, R. Harris, and E. Hultman, "Resynthesis of Creatine Phosphate in Human Muscle after Exercise in Relation to Intramuscular pH and Availability of Oxygen," *Scandinavian Journal of Clinical and Laboratory Investigation* 39, no. 6 (October 1979): 551–58, https://doi.org/10.3109/00365517909108833.

65. McArdle, Katch, and Katch, *Essentials*, 185.

66. Paul Fournier et al., "Post-Exercise Muscle Glycogen Repletion in the Extreme: Effect of Food Absence and Active Recovery," *Journal of Sports Science and Medicine* 3 (2004): 139–46, https://doi.org/10.0000/www.ncbi.nlm.nih.gov/PMC3905296.

67. Brooks, "Cell-Cell."

68. Daniela Valenti et al., "Lactate Transport into Rat Heart Mitochondria and Reconstruction of the l-Lactate/Pyruvate Shuttle," *Biochemical Journal* 364, no. 1 (2002): 101–4, https://doi.org/10.1042/bj3640101.

69. Carsten Juel et al., "Effect of High-Intensity Intermittent Training on Lactate and H+ Release from Human Skeletal Muscle," *American Journal of Physiology-Endocrinology and Metabolism* 286, no. 2 (February 2004): E245–51, https://doi.org/10.1152/ajpendo.00303.2003.

70. Mauro Sola-Penna, "Metabolic Regulation by Lactate," *IUBMB Life* 60, no. 9 (2008): 605–8, https://doi.org/10.1002/iub.97.

71. Kristin Osterling et al., "The Effects of High Intensity Exercise during Pulmonary Rehabilitation on Ventilatory Parameters in People with Moderate to Severe Stable COPD: A Systematic Review," *International Journal of Chronic Obstructive Pulmonary Disease* (2014): 1069, https://doi.org/10.2147/copd.s68011.

72. Boutcher, "High-Intensity Intermittent Exercise."

73. Hai-Jun Xu, "Role of Lactate in Lipid Metabolism, Just Always Inhibiting Lipolysis?" *Journal of Biological Chemistry* 284, no. 31 (June 2009): le5, https://doi.org/10.1074/jbc.L806409200.

74. Jared Fletcher and Brian MacIntosh, "Running Economy from a Muscle Energetics Perspective," *Frontiers in Physiology* 8, no. 433 (June 2017): 1–15, https://doi.org/10.3389/fphys.2017.00433.

75. Trapp et al., "The Effects of High-Intensity."

76. McArdle, Katch, and Katch, *Essentials*, 156.

77. Noriyuki Ouchi, Rei Shibata, and Kenneth Walsh, "AMP-Activated Protein Kinase Signaling Stimulates VEGF Expression and Angiogenesis in Skeletal Muscle," *Circulation Research* 96, no. 8 (April 2005): 838–46, https://doi.org/10.1161/01.res.0000163633.10240.3b.

78. Paul Laursen, "Training for Intense Exercise Performance: High-Intensity or High-Volume Training?" *Scandinavian Journal of Medicine and Science in Sports* 20 (2010): 1–10, https://doi.org/10.1111/j.1600-0838.2010.01184.x.

79. Laursen, "Training for Intense."

80. Laursen, "Training for Intense."

81. Su Myung Jung, Joan Sanchez-Gurmaches, and David A. Guertin, "Brown Adipose Tissue Development and Metabolism," *Handbook of Experimental Pharmacology* 251 (2018): 3–36, http://doi.org/10.1007/164_2018_168.

82. Jung, Sanchez-Gurmaches, and Guertin, "Brown Adipose"; Rebecca Tunstall, "Exercise Training Increases Lipid Metabolism Gene Expression in Human Skeletal Muscle," *American Journal of Physiology-Endocrinology and Metabolism* 283, no.1 (July 2002): E66–72, https://doi.org/10.1152/ajpendo.00475.2001.

83. Paul Lee et al., "Irisin and FGF21 Are Cold-Induced Endocrine Activators of Brown Fat Function in Humans," *Cell Metabolism* 19, no. 2 (February 2014): 302–9, https://doi.org/10.1016/j.cmet.2013.12.017.

84. Lee et al., "Irisin."

85. Brooks, "Cell-Cell."

86. Rafael Gaspar et al., "Acute Physical Exercise Increases Leptin-Induced Hypothalamic Extracellular Signal-Regulated Kinase1/2 Phosphorylation and Thermogenesis of Obese Mice," *Journal of Cellular Biochemistry* (2018): 1–8, https://doi.org/10.1002/jcb.27426.

87. Jung, Sanchez-Gurmaches, and Guertin, "Brown Adipose."

88. Fletcher and MacIntosh, "Running Economy."

89. Jeffrey Horowitz and Samuel Klein, "Lipid Metabolism and Endurance Exercise," *American Journal of Clinical Nutrition* 72, no. 2 (2000): 558S–63S, https://doi.org/10.1093/ajcn/72.2.558s; Darlene Sedlock, Jean Fissinger, and Christopher Melby, "Effect of Exercise Intensity and Duration on Postexercise Energy Expenditure," *Medicine and Science in Sports and Exercise* 21, no. 6 (1989): 662–66, https://doi.org/10.1249/00005768-198912000-00006.

90. Fletcher and MacIntosh, "Running Economy."

91. J. B. Warren et al., "Adrenaline Secretion during Exercise," *Clinical Science* 65, no. 3 (January 1984): 87–90, https://doi.org/10.1042/cs065027pb.

92. McArdle, Katch, and Katch, *Essentials*, 372.

Chapter 8: Food Basics

1. Ronghua Zhang et al., "The Difference in Nutrient Intakes between Chinese and Mediterranean, Japanese, and American Diets," *Nutrients* 7, no. 6 (2015): 4661–68, https://doi.org/10.3390/nu7064661.

2. Kayo Kurotani et al., "Quality of Diet and Mortality among Japanese Men and Women: Japan Public Health Center Based Prospective Study," *British Medical Journal* 352 (2016): i1209, https://doi.org/10.1136/bmj.i1209.

3. Christopher Gardner et al., "Effect of Low-Fat vs. Low-Carbohydrate Diet on 12-Month Weight Loss in Overweight Adults and the Association with Genotype Pattern or Insulin Secretion," *Journal of the American Medical Association* 319, no. 7 (February 2018): 667–79, https://doi.org/10.1001/jama.2018.0245.

4. Fung, *Obesity Code*.

5. Ouyang Xiaosen et al., "Fructose Consumption as a Risk Factor for Non-Alcoholic Fatty Liver Disease," *Journal of Hepatology* 48, no. 6 (June 2008): 993–99, https://doi.org/10.1016/j.jhep.2008.02.011.

6. Prasanthi Jegatheesan and Jean-Pascal De Bandt, "Fructose and NAFLD: The Multifaceted Aspects of Fructose Metabolism," *Nutrients* 9, no. 3 (March 2017): E230, https://doi.org/10.3390/nu9030230; Xiaosen et al., "Fructose Consumption."

7. Miguel Baena et al., "Fructose, but Not Glucose, Impairs Insulin Signaling in the Three Major Insulin-Sensitive Tissues," *Scientific Reports* 6, no. 1 (May 2016): 26149, https://doi.org/10.1038/srep26149.

8. Adam Drewnowski and Colin Rehm, "Consumption of Added Sugars Among US Children and Adults by Food Purchase Location and Food Source," *American Journal of Clinical Nutrition* 100, no. 3 (September 2014): 901–7, https://doi.org/10.3945/ajcn.114.089458.
9. Drewnowski and Rehm, "Consumption."
10. You Xu et al., "Prevalence and Control of Diabetes in Chinese Adults," *Journal of the American Medical Association* 310, no. 9 (2013): 948–59, https://doi.org/10.1001/jama.2013.168118.
11. S. H. Ahmed, K. Guillem, and Y. Vandaele, "Sugar Addiction: Pushing the Drug-Sugar Analogy to the Limit," *Current Opinion in Clinical Nutrition and Metabolic Care* 16, no. 4 (July 2013): 434–39, https://doi.org/ 10.1097/MCO.0b013e328361c8b8.
12. Douglas Paddon-Jones et al., "Role of Dietary Protein in the Sarcopenia of Aging," *American Journal of Clinical Nutrition* 87, no. 5 (May 2008): 1562S–26S, https://doi.org/10.1093/ajcn/87.5.1562s; Anssi Manning, "Very-Low-Carbohydrate Diets and Preservation of Muscle Mass," *Nutrition and Metabolism* 3, no. 1 (2006): 9, https://doi.org/10.1186/1743-7075-3-9.
13. McArdle, Katch, and Katch, *Essentials.*
14. Denise Robertson et al., "Insulin-Sensitizing Effects of Dietary Resistant Starch and Effects on Skeletal Muscle and Adipose Tissue Metabolism," *American Journal of Clinical Nutrition* 82, no. 3 (September 2005): 559–67, https://doi.org/10.1093/ajcn/82.3.559.
15. Tracy Horton et al., "Fat and Carbohydrate Overfeeding in Humans: Different Effects on Energy Storage," *American Journal of Clinical Nutrition* 62, no. 1 (July 1995): 19–29, https://doi.org/10.1093/ajcn/62.1.19.
16. A. P. Simopoulos, "The Importance of the Ratio of Omega-6/Omega-3 Essential Fatty Acids," *Biomedical Pharmacotherapy* 56, no. 8 (October 2002): 365–79, https://doi.org/10.1016/s0753-3322(02)00253-6.
17. Price Schönfeld and Lech Wojtczak, "Short- and Medium-Chain Fatty Acids in Energy Metabolism: The Cellular Perspective," *Journal of Lipid Research* 57, no. 6 (June 2016): 943–54, https://doi.org/10.1194/jlr.r067629.
18. Ming-Hua Sung, Fang-Hsuean Liao, and Yi-Wen Chien, "Medium-Chain Triglycerides Lower Blood Lipids and Body Weight in Streptozotocin-Induced Type 2 Diabetes Rats," *Nutrients* 10, no. 8 (July 2018): E963, https://doi.org/10.3390/nu10080963.
19. W. Fernando et al., "The Role of Dietary Coconut for the Prevention and Treatment of Alzheimer's Disease: Potential Mechanisms of Action," *British Journal of Nutrition* 114, no. 1 (July 2015): 1–14, https://doi.org/10.1017/s0007114515001452.
20. Lee Know, *Mitochondria and the Future of Medicine* (White River Junction, VT: Chelsea Green Publishing, 2018), 139–66.
21. Frank Sacks et al., "Comparison of Weight-Loss Diets with Different Compositions of Fat, Protein, and Carbohydrates," *New England Journal of Medicine* 360, no. 9 (February 2009): 859–73, https://doi.org/10.1056/nejmoa0804748.
22. Amy Berger, *The Alzheimer's Antidote: Using a Low-Carb, High-Fat Diet to Fight Alzheimer's Disease, Memory Loss, and Cognitive Decline* (White River Junction, VT: Chelsea Green Publishing, 2017).

Chapter 9: Managing Diets and Fastercise

1. Jeff Volek and Stephen Phinney, *The Art and Science of Low Carbohydrate Living* (Florida: Beyond Obesity, 2011).
2. Xiaodong Zhuang et al., "U-Shaped Relationship between Carbohydrate Intake Proportion and Incident Atrial Fibrillation," *Journal of the American College of Cardiology* 73, no. 9, sup. 2 (March 2019): 4, https://doi.org/10.1016/S0735-1097(19)33766-0.
3. Jeff Volek and Stephen Phinney, *The Art and Science of Low Carbohydrate Performance* (Florida: Beyond Obesity, 2012), 79.
4. Supornpim Chearskul et al., "Effect of Weight Loss and Ketosis on Postprandial Cholecystokinin and Free Fatty Acid Concentrations," *American Journal of Clinical Nutrition* 87, no. 5 (May 2008): 1238–46, https://doi.org/10.1093/ajcn/87.5.1238.
5. Jørn Helge, "Long-Term Fat Diet Adaptation Effects on Performance, Training Capacity, and Fat Utilization," *Medicine and Science in Sports and Exercise* 34, no. 9 (September 2002): 1499–504, https://doi.org/10.1249/01.MSS.0000027691.95769.B5.
6. Volek and Phinney, *Low Carbohydrate Performance*.
7. Krista Howarth et al., "Effect of Glycogen Availability on Human Skeletal Muscle Protein Turnover during Exercise and Recovery," *Journal of Applied Physiology* 109, no. 2 (August 2010): 431–38, https://doi.org/10.1152/japplphysiol.00108.2009.
8. Kevin Acheson et al., "Glycogen Storage Capacity and de novo Lipogenesis during Massive Carbohydrate Overfeeding in Man," *American Journal of Clinical Nutrition* 48, no. 2 (1988): 240–47, https://doi.org/10.1093/ajcn/48.2.240.
9. Shivani Sahni et al., "Higher Protein Intake Is Associated with Higher Lean Mass and Quadriceps Muscle Strength in Adult Men and Women," *Journal of Nutrition* 145, no. 7 (July 2015): 1569–75, https://doi.org/10.3945/jn.114.204925.
10. James Krieger et al., "Effects of Variation in Protein and Carbohydrate Intake on Body Mass and Composition during Energy Restriction: A Meta-Regression," *American Journal of Clinical Nutrition* 83, no. 2 (2006): 260–74, https://doi.org/10.1093/ajcn/83.2.260.
11. Mastaneh Sharafi et al., "Effect of a High-Protein, High-Fiber Beverage Preload on Subjective Appetite Ratings and Subsequent ad Libitum Energy Intake in Overweight Men and Women: A Randomized, Double-Blind Placebo-Controlled, Crossover Study," *Current Developments in Nutrition* 2, no. 6 (June 2018): nzy022, https://doi.org/10.1093/cdn/nzy022; Thomas Halton and Frank Hu, "The Effects of High Protein Diets on Thermogenesis, Satiety and Weight Loss: A Critical Review," *Journal of the American College of Nutrition* 23, no. 5 (October 2004): 373–85, https://doi.org/10.1080/07315724.2004.10719381.
12. P. Clifton, K. Bastiaans, and J. Keogh, "High Protein Diets Decrease Total and Abdominal Fat and Improve CVD Risk Profile in Overweight and Obese Men and Women with Elevated Triacylglycerol," *Nutrition, Metabolism and Cardiovascular Disease* 19, no. 8 (October 2009): 548–54, https://doi.org/10.1016/j.numecd.2008.10.006.
13. Acheson et al., "Glycogen Storage."

14. Acheson et al., "Glycogen Storage."
15. Acheson et al., "Glycogen Storage."
16. Acheson et al., "Glycogen Storage."
17. Acheson et al., "Glycogen Storage."
18. Roy Jentjens and Asker Jeukendrup, "Determinants of Post-Exercise Glycogen Synthesis during Short-Term Recovery," *Sports Medicine* 33, no. 2 (2003): 117–44, https://doi.org/10.2165/00007256-200333020-00004.
19. Jentjens and Jeukendrup, "Determinants."
20. Jeff Volek et al., "Metabolic Characteristics."
21. Jeff Volek, Erin Quann, and Cassandra Forsythe, "Low-Carbohydrate Diets Promote a More Favorable Body Composition Than Low-Fat Diets," *Strength and Conditioning Journal* 32, no. 1 (2010): 42–47, https://doi.org/10.1519/ssc.0b013 e3181c16c41.
22. Volek and Phinney, *Low Carbohydrate Performance*, 7.
23. Brooks, "Cell-Cell."
24. Volek and Phinney, *Low Carbohydrate Performance*.
25. Kiyotaka Sato et al., "Insulin, Ketone Bodies, and Mitochondrial Energy Transduction," *FASEB Journal* 9, no. 8 (May 1995): 651–58, https://doi.org /10.1096/fasebj.9.8.7768357; Richard Veech, "The Therapeutic Implications of Ketone Bodies: The Effects of Ketone Bodies in Pathological Conditions: Ketosis, Ketogenic Diet, Redox States, Insulin Resistance, and Mitochondrial Metabolism," *Prostaglandins, Leukotrienes and Essential Fatty Acids* 70, no. 3 (2004): 309–19, https:// doi.org/10.1016/j.plefa.2003.09.007.
26. B. Glancy et al., "Effect of Calcium on the Oxidative Phosphorylation Cascade in Skeletal Muscle Mitochondria," *Biochemistry* 52, no. 16 (April 2013): 2793–809, https://doi.org/ 10.1021/bi3015983.
27. George Brooks et al., "Cardiac and Skeletal Muscle Mitochondria Have a Monocarboxylate Transporter MCT1," *Journal of Applied Physiology* 87, no. 5 (1999): 1713–18, https://doi.org/10.1152/jappl.1999.87.5.1713.
28. Daniel Kane, "Lactate Oxidation at the Mitochondria: A Lactate-Malate-Aspartate Shuttle at Work," *Frontiers in Neuroscience* 8 (November 2014): 366, https://doi.org /10.3389/fnins.2014.00366.
29. U. Keller et al., "Effects of Changes in Hydration on Protein, Glucose and Lipid Metabolism in Man: Impact on Health," *European Journal of Clinical Nutrition* 57, supplement 2 (December 2003): S69–74, https://doi.org/10.1038/1601904.
30. Jeff Volek et al., "Metabolic Characteristics"; Mark Evans, Karl Cogan, and Brendan Egan, "Metabolism of Ketone Bodies during Exercise and Training: Physiological Basis for Exogenous Supplementation," *Journal of Physiology* 595, no. 9 (2016): 2857–71, https://doi.org/10.1113/jp273185.
31. Andrew Coggan et al., "Fat Metabolism during High-Intensity Exercise in Endurance-Trained and Untrained Men," *Metabolism* 49, no. 1 (2000): 122–28, https://doi.org/10.1016/s0026-0495(00)90963-6.

32. Volek and Phinney, *Low Carbohydrate Performance*, 14.
33. Pete Cox et al., "Nutritional Ketosis Alters Fuel Preference and Thereby Endurance Performance in Athletes," *Cell Metabolism* 24, no. 2 (2016): 256–68, https://doi.org /10.1016/j.cmet.2016.07.010.
34. R. Passmore and R. Johnson, "The Modification of Post-Exercise Ketosis (the Courtice-Douglas Effect) by Environmental Temperature and Water Balance," *Quarterly Journal of Experimental Physiology and Cognate Medical Sciences* 43, no. 4 (1958): 352–61, https://doi.org/10.1113/expphysiol.1958.sp001348.
35. Peter Wilde, "Eating for Life: Designing Foods for Appetite Control," *Journal of Diabetes Science and Technology* 3, no. 2 (2009): 366–70, https://doi.org/10.1177 /193229680900300219.
36. R. Johnson and J. Walton, "Fitness, Fatness, and Post-Exercise Ketosis," *The Lancet* 297, no. 7699 (1971): 566–68, https://doi.org/10.1016/s0140-6736(71)91164-0.
37. Evans, Cogan, and Egan, "Metabolism of Ketone Bodies."
38. Françoise Féry and Edmond Balasse, "Ketone Body Turnover during and after Exercise in Overnight-Fasted and Starved Humans," *American Journal of Physiology-Endocrinology and Metabolism* 245, no. 4 (1983): E318–25, https://doi.org/10.1152 /ajpendo.1983.245.4.e318.
39. Volek et al., "Metabolic Characteristics."
40. Volek and Phinney, *Low Carbohydrate Performance*, 33.
41. Volek et al., "Metabolic Characteristics."
42. Kaito Iwayama et al., "Exercise before Breakfast Increases 24-h Fat Oxidation in Female Subjects," *PLoS ONE* 12, no. 7 (July 2017): e0180472, https://doi.org/10 .1371/journal.pone.0180472.

Chapter 10: The Promise of Fastercise

1. Victor R. Preedy, Gerard N. Burrow, and Ronald Watson, eds., *Comprehensive Handbook of Iodine* (Amsterdam: Elsevier, 2009): chapter 105, https://www.sciencedirect.com /science/article/pii/B9780123741356001059.
2. James Gillooly et al., "Effects of Size and Temperature on Metabolic Rate," *Science* 293 (September 2001): 2248–51, https://doi.org/10.1126/science.1061967.
3. Yamini Tripathi, O. Malhotra, and S. Tripathi, "Thyroid Stimulating Action of Z-Guggulsterone Obtained from *Commiphora mukul*," *Planta Medica* 50, no. 1 (1984): 78–80, https://doi.org/10.1055/s-2007-969626; Sunanda Panda and Anand Kar, "Gugulu (*Commiphora mukul*) Induces Triiodothyronine Production: Possible Involvement of Lipid Peroxidation," *Life Sciences* 62, no. 12 (1999): PL137–41, https://doi.org/10.1016/s0024-3205(99)00369-0.
4. Silva and Bianco, "Thyroid–Adrenergic."
5. Winifred Yau et al., "Thyroid Hormone (T3) Stimulates Brown Adipose Tissue Activation via Mitochondrial Biogenesis and MTOR-Mediated Mitophagy," *Autophagy* 15, no. 1 (2019): 1–20, https://doi.org/10.1080/15548627.2018.1511263.

INDEX

Note: Page numbers followed by "f" refer to figures. Page numbers followed by "t" refer to tables.

ABOUT THE AUTHOR

Denis Wilson, MD, is the author of *Wilson's Temperature Syndrome, Doctor's Manual for Wilson's Temperature Syndrome,* and *Evidence-Based Approach to Restoring Thyroid Health.* As the originator of the WT3 protocol, he has been educating physicians for over twenty-six years on the use of sustained-release T3 in the treatment of Wilson's temperature syndrome, a condition in which people exhibit low thyroid and low body temperature symptoms but have normal thyroid blood test results. Dr. Wilson speaks at medical conventions and medical schools both nationally and internationally and trains physicians on the use of herbs and nutrients.